Undergraduate Lecture Notes in Physics

W0079138

For further volumes:
http://www.springer.com/series/8917

Undergraduate Lecture Notes in Physics (ULNP) publishes authoritative texts covering topics throughout pure and applied physics. Each title in the series is suitable as a basis for undergraduate instruction, typically containing practice problems, worked examples, chapter summaries, and suggestions for further reading.

ULNP titles must provide at least one of the following:

- An exceptionally clear and concise treatment of a standard undergraduate subject.
- A solid undergraduate-level introduction to a graduate, advanced, or non-standard subject.
- A novel perspective or an unusual approach to teaching a subject.

ULNP especially encourages new, original, and idiosyncratic approaches to physics teaching at the undergraduate level.

The purpose of ULNP is to provide intriguing, absorbing books that will continue to be the reader's preferred reference throughout their academic career.

Series Editors

Neil Ashby
Professor Emeritus, University of Colorado, Boulder, CO, USA

William Brantley
Professor, Furman University, Greenville, SC, USA

Michael Fowler
Professor, University of Virginia, Charlottesville, VA, USA

Michael Inglis
Professor, SUNY Suffolk County Community College, Selden, NY, USA

Heinz Klose
Oldenburg, Niedersachsen, Germany

Helmy Sherif
Professor Emeritus, University of Alberta, Edmonton, AB, Canada

Walter J. Maciel

Hydrodynamics and Stellar Winds

An Introduction

 Springer

Walter J. Maciel
Departamento de Astronomia
Cidade Universitaria IAG/USP
São Paulo, SP, Brazil

Translation from the Portuguese language edition: *Hidrodinâmica e Ventos Estelares: Uma Introdução*, by Walter J. Maciel, © Editora da Universidade de São Paulo 2005. All rights reserved.

ISSN 2192-4791   ISSN 2192-4805 (electronic)
ISBN 978-3-319-04327-2   ISBN 978-3-319-04328-9 (eBook)
DOI 10.1007/978-3-319-04328-9
Springer Cham Heidelberg New York Dordrecht London

Library of Congress Control Number: 2014931572

Printed on acid-free paper

Springer is part of Springer Science+Business Media (www.springer.com)

Così tra questa immensità s'annega il pensier mio:
E il naufragar m'è dolce in questo mare.

Leopardi, L'infinito

For Juca and Nicota, now reunited

Preface

Stellar winds are a common phenomenon in stars, from the dwarfs, like the Sun, to the hot supergiants, and including red giants with low surface temperatures. In fact, we can say that all stars lose mass during some periods of their lives, and the study of this phenomenon is one of the main aspects of modern astrophysics. In the case of the Sun and other dwarf stars, the mass loss rate is small, having a small effect on the stellar evolution tracks, but even in these cases other effects may be important, such as the influence of the solar wind on the communications in our planet. In the case of hot stars, or even for red giants, the observed rates can be significant, thus affecting sometimes drastically the evolution and outcome of these stars.

Stellar winds are hydrodynamic phenomena involving the flow of circumstellar gases, so that the basic hydrodynamic concepts are directly applied in the study of the winds and of the mass loss from the stars. Unfortunately, many students in the astrophysical courses are not familiar with the foundations of this discipline, which have to be studied at the same time as they are introduced to the observational evidences of stellar winds, theoretical models and especially the association of purely hydrodynamical phenomena with the radiative transfer in the outer atmospheres of the stars. In this way, the study of the stellar winds becomes excessive, which frequently limits the amount of material that can be taught in the courses on the winds or on stellar physics.

This little book aims at partially compensating for the lack of basic knowledge of the hydrodynamic principles of astrophysics students. It is not a complete course on hydrodynamics, its astrophysical applications and stellar winds, but just an introduction to these important aspects of stellar physics. The text originated from lecture notes of a graduate course on stellar winds presented in the last 20 years at the astronomy department of São Paulo University.

The book attempts to present the main basic hydrodynamics equations, with many examples and suggested exercises. In this way, it can also be used as an introductory hydrodynamics text for physics and related students, although its main goal – and the applications worked out – are essentially dedicated to stellar astrophysics. The basic principles of hydrodynamics have many astrophysical applications, as in gas dynamics or in the dynamics of galaxies and galaxy clusters, so that the book is particularly useful for students of general astrophysics. Exercises and a basic bibliography are included at the end of each chapter, allowing a more complete understanding of the subjects treated in the book.

Some parts of the book were inspired by graduate courses given at IAG/USP along the years, especially those by professor J. A. de Freitas Pacheco, to whom I am indebted. Preliminary versions of the text were supported by the department chairpersons Beatriz Barbuy, Jacques Lépine, and Vera Jatenco. I am also indebted to Augusto Damineli for the spectrum of de P Cyg, to Rodrigo Campos for the images of the Sun, M27 and the Crab Nebula, to Graziela Keller, for the spectra of NGC 6905, and to all colleagues and students that have collaborated in the revision of previous versions of this book, particularly Vera Jatenco, for the many corrections and suggestions. Naturally, any remaining mistakes and omissions are of my own responsibility.

The present edition is a translation of the book originally published in 2005. A few mistakes have been corrected, the bibliography was updated, and several new examples and applications have been included, but the main goals of the original edition have been preserved.

São Paulo, Brazil   Walter J. Maciel

About the Author

Walter Junqueira Maciel was born in Cruzília, MG, Brazil. He graduated in Physics at UFMG (Minas Gerais Federal University), in Belo Horizonte, and obtained a master's degree at ITA (Aeronautics Technological Institute), São José dos Campos, and a PhD at São Paulo University. He did several internships in Groningen, the Netherlands, and in Heidelberg, Germany. He is a full professor in the Astronomy Department at the Astronomical and Geophysical Institute, São Paulo University, where he has been working since 1974. He was the head of department between 1992 and 1994. He has published over a hundred scientific papers in international journals and around 50 papers (science, education, and outreach) in national journals. He is the author of the book "Introduction to Stellar Structure and Evolution" (Edusp, 1999), which won the Jabuti Prize in 2000 in the field of exact sciences, technology, and computer sciences, and of the book Astrophysics of the Interstellar Medium (Springer, 2013).

Contents

Chapter 1
The Continuity Equation

Abstract This chapter discusses the continuum hypothesis, and presents the continuity equation, which is written in spherical and cylindrical coordinates. The chapter ends with a discussion of the mass loss rate as related to the continuity equation. Several examples are worked out, involving the loss of mass in hot and cool stars.

1.1 Introduction

Most stars lose mass to the interstellar medium during their evolution. Mass loss is particularly intense in hot stars, such as main sequence stars with early spectral types, hot supergiants, red giants, Wolf-Rayet stars and central stars of planetary nebulae. The mechanisms responsible for the mass ejection depend on the kind of stars, and are not always well known.

In this book, we will be basically interested in the process of mass ejection by stars in the form of stellar winds. This process is extremely complex in its general form, as it involves the coupling of the hydrodynamic flow with the transfer of the stellar emergent radiation. Therefore, it is convenient to review the main hydrodynamics concepts, aiming at their application to the study of the stellar winds. We will initially consider the main equations that characterize the hydrodynamic flows, and then analyze a few simple cases of mass loss in stellar winds.

The basic reference to the study of stellar winds is the work by Lamers and Cassinelli (1999). A rigorous discussion of the physics of stellar atmospheres and of the process of mass loss can also be found in Mihalas (1978). Basic hydrodynamics concepts can be found in many publications, such as Landau and Lifchitz (1971) and Batchelor (1967).

W.J. Maciel, *Hydrodynamics and Stellar Winds: An Introduction*, Undergraduate Lecture Notes in Physics, DOI 10.1007/978-3-319-04328-9_1,
© Springer International Publishing Switzerland 2014

1.2   The Continuum Hypothesis

Let us consider a gas kept at constant temperature in an enclosed container. We generally associate the *pressure* by the gas on the container walls to the force per unit area communicated to the walls by the gas particles through elastic collisions. Since the gas is at a constant temperature and the container is closed, the pressure will be constant in the walls.

Let us now consider the same container with a very small quantity of gas, that is, just a few molecules. In this case, the pressure as defined above is not a *continuous* property of the gas, and will depend on the collision probabilities of the molecules with the walls. This example clarifies the so-called *continuum hypothesis*, that is, the conditions that must be fulfilled so that the system can be considered as a fluid. That system can be the gas in an stellar envelope or a stellar cluster.

Generally, a substance is considered as a fluid if the smallest volume element to be considered contains a sufficient number of particles (atoms, molecules, stars, etc.) in order that its average properties vary in a continuous way. In other words, what we call a *point* or *volume element* in a fluid may contain in fact a large number of particles.

The continuum hypothesis can also be characterized in terms of the mean free path of the fluid particles. In fact, the mean free path of a particle is inversely proportional to the number density of the particles of the considered substance. Therefore, in order that a given volume element be considered as continuous, the molecular density must be sufficiently high, or the mean free path λ must be small with relation to some characteristic dimension R of the system, that is

$$\lambda \ll R. \tag{1.1}$$

1.2.1   Example: Glass of Water

Let us consider a simple example of a fluid: a glass containing water. Considering the density of water in the usual conditions $\rho \simeq 1\,\mathrm{g/cm^3}$ and the molecular mass $m \simeq 18\,m_H$, where $m_H = 1.67 \times 10^{-24}$ g is the mass of a hydrogen atom, the number density of the water molecules in the glass is

$$n \simeq \frac{\rho}{m} \simeq \frac{\rho}{18\,m_H} \simeq 3 \times 10^{22}\ \mathrm{cm^{-3}}.$$

In order of magnitude, the molecular mean free path can be approximated by the average separation of the molecules,

$$\lambda \sim n^{-1/3} \simeq 3 \times 10^{-8}\ \mathrm{cm}.$$

A characteristic dimension of the system (the glass) is the average radius, $R \simeq 5\,\mathrm{cm}$. Therefore, we have $\lambda \ll R$, so that the continuum hypothesis is fulfilled.

1.2.2   Example: Grains in a Stellar Envelope

Red giant stars lose mass to the interstellar medium in a continuous way, so that they have a circumstellar envelope beyond their photospheric layers. Dust grains are located in the envelope along with the gas particles, and are probably responsible for at least part of the mass loss. Let us consider as a second example the circumstellar envelope of a red giant star, where both solid grains and gas coexist. The number density of the gas particles is typically $n \sim 10^8$ cm^{-3}. We can consider the grains as spherical particles with an average radius $a \simeq 1,000$ Å $= 10^{-5}$ cm. The cross section for grain-gas collisions, taken as the grain geometrical cross section, is given by

$$\sigma \sim \pi a^2 \sim 3 \times 10^{-10} \text{ cm}^2.$$

Therefore, the grain mean free path through the gas is

$$\lambda \sim \frac{1}{n\,\sigma} \sim 30 \text{ cm}.$$

A characteristic dimension of a circumstellar envelope is its radius, which is typically of the order of several stellar radii, or $R \sim 10^{14}$ cm. We see that $\lambda \ll R$, so that we can treat the envelope containing gas and grains as a fluid. We can also estimate the mean free path for collisions among the gas particles themselves. We have

$$\lambda_g \sim n^{-1/3} \simeq 2 \times 10^{-3} \text{ cm}.$$

which is even smaller than the mean free path for grain-gas collisions.

1.3   The Continuity Equation

The state of an ideal fluid can be described by the velocity distribution of the fluid, characterized by the vector $\mathbf{v}(x, y, z, t)$ and by two thermodynamic variables, such as the pressure $P(x, y, z, t)$ and the density $\rho(x, y, z, t)$. Vector $\mathbf{v}$ describes the fluid velocity at a point in space characterized by the cartesian coordinates (x, y, z) at time t, and the same is valid for P and ρ. Let us consider a volume V of the fluid. The total fluid mass contained in this volume is

$$\int_V \rho \, dV,$$

and can be measured in grams (g). The integral extends to the whole volume V. It should be noted that a fluid volume element may contain a large number of molecules, although it is small compared with other dimensions involved, such as the scale height in a stellar envelope.

Fig. 1.1 Characterization of
a surface element dS in a fluid
with volume V

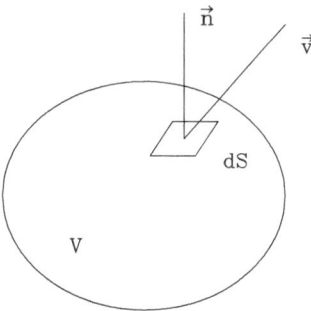

Let dS be the surface element of the area that limits the considered volume, with
a unit vector $\mathbf{n}$ (Fig. 1.1). Vector $\mathbf{n}$ is perpendicular to the surface element dS and
directed outwards of the element. The amount of matter that crosses dS per unit
time is

$$\rho\,\mathbf{v}\cdot\mathbf{n}\,dS,$$

and can be measured for example in $(\text{g/cm}^3)\,(\text{cm/s})\,(\text{cm}^2) = \text{g/s}$. Therefore, $\rho\,\mathbf{v}\cdot\mathbf{n}\,dS$ is positive when the fluid *leaves* the volume element and negative when it
enters the considered volume. The total fluid mass that leaves volume V per unit
time is

$$\oint \rho\,\mathbf{v}\cdot\mathbf{n}\,dS,$$

given in g/s. The integral extends through the whole surface around volume V. On
the other hand, we may write that the fluid mass decrease per unit time is

$$-\frac{\partial}{\partial t}\int_V \rho\,dV,$$

given in $(\text{s}^{-1})(\text{g/cm}^3)(\text{cm}^3) = \text{g/s}$. The negative sign $(-)$ is due to the fact that the
amount of matter in volume V decreases. In order to have mass conservation, we
must have

$$\frac{\partial}{\partial t}\int_V \rho\,dV = -\oint \rho\,\mathbf{v}\cdot\mathbf{n}\,dS. \tag{1.2}$$

According to the divergence theorem, if $\mathbf{A}$ is a vector,

$$\int_V \boldsymbol{\nabla}\cdot\mathbf{A}\,dV = \oint \mathbf{A}\cdot\mathbf{n}\,dS, \tag{1.3}$$

so that

$$\oint \rho\,\mathbf{v}\cdot\mathbf{n}\,dS = \int_V \boldsymbol{\nabla}\cdot(\rho\,\mathbf{v})\,dV. \tag{1.4}$$

From Eqs. (1.2) and (1.4),

$$\frac{\partial}{\partial t} \int_V \rho \, dV = - \int_V \nabla \cdot (\rho \, \mathbf{v}) \, dV$$

or

$$\int_V \left[\frac{\partial \rho}{\partial t} + \nabla \cdot (\rho \, \mathbf{v}) \right] dV = 0. \qquad (1.5)$$

The equality above must hold for any chosen volume. Therefore, we can write

$$\frac{\partial \rho}{\partial t} + \nabla \cdot (\rho \, \mathbf{v}) = 0, \qquad (1.6)$$

with units $\mathrm{g\,cm^{-3}\,s^{-1}}$. This is the *mass conservation equation*, or *continuity equation*. Since

$$\nabla \cdot (\rho \, \mathbf{v}) = \rho \, \nabla \cdot \mathbf{v} + \mathbf{v} \cdot \nabla \rho, \qquad (1.7)$$

Eq. (1.6) can be written as

$$\frac{\partial \rho}{\partial t} + \rho \, \nabla \cdot \mathbf{v} + \mathbf{v} \cdot \nabla \rho = 0. \qquad (1.8)$$

Naturally, the continuity equation (1.6) or (1.8) expresses the mass conservation. In the absence of sources or sinks, the amount of matter leaving a certain volume of the fluid is the same as the mass reduction in the same volume. In other words, the density in a certain volume element varies according to the flow in or out of the element.

1.3.1   Example: Incompressible Fluids

A particular case of Eq. (1.6) is that of incompressible fluids, for which the density ρ can be taken as constant. In this case, the continuity equation is simply

$$\nabla \cdot \mathbf{v} = 0. \qquad (1.9)$$

1.3.2   Example: Steady State

An important case is the *steady state*, in which all flow properties are constant with time. The term $\partial \rho / \partial t = 0$, and Eq. (1.6) is written as

$$\nabla \cdot (\rho \, \mathbf{v}) = 0. \qquad (1.10)$$

1.4   The Mass Flux

Let us consider the quantity defined as

$$\mathbf{j} = \rho\,\mathbf{v} \tag{1.11}$$

with units: $(g/cm^3)(cm/s) = g\,cm^{-2}\,s^{-1}$. This vector has the same direction as vector $\mathbf{v}$, and can be clarified by Eq. (1.2). The integral $\int \rho\,dV$ in this equation gives the total mass in volume V, so that the first member of the equation gives the rate of change of this mass with time. The second member takes into account the mass that traverses surface S around the volume considered per unit time, which is the integral of vector $\rho\,\mathbf{v}$ in the whole surface S. In this case, vector $\mathbf{j} = \rho\,\mathbf{v}$ can be identified with the *mass flux* through surface S, that is, it corresponds to the amount of matter that traverses a unit surface element perpendicular to $\mathbf{v}$ per unit time. We will see later that, analogously to Eqs. (1.2) and (1.11), we may define an *energy flux* and a *momentum flux*. In the present case, the mass flux is a *vector*, and the mass is a *scalar* quantity.

In terms of the mass flux $\mathbf{j}$, the continuity equation (1.6) can be written as

$$\frac{\partial \rho}{\partial t} + \nabla \cdot \mathbf{j} = 0, \tag{1.12}$$

which is also found in electromagnetism, where it expresses the conservation of electric charge (taking ρ as the charge density and $\mathbf{j}$ as the electric current).

1.5   Spherical Coordinates

In this book we will be especially interested in problems with spherical symmetry, where a central star ejects mass into the interstellar medium. Using the spherical coordinates r, θ, ϕ (Fig. 1.2), we have the following relations:

$$\begin{cases} x = r\,\sin\theta\,\cos\phi \\ y = r\,\sin\theta\,\sin\phi \\ z = r\,\cos\theta. \end{cases} \tag{1.13}$$

The volume element dV can be written as:

$$dV = r^2\,\sin\theta\,dr\,d\theta\,d\phi. \tag{1.14}$$

Taking $\mathbf{A}$ as a vector with coordinates (A_r, A_θ, A_ϕ), we have

$$\nabla \cdot \mathbf{A} = \frac{1}{r^2}\frac{\partial}{\partial r}\left(r^2\,A_r\right) + \frac{1}{r\,\sin\theta}\frac{\partial}{\partial \theta}\left(\sin\theta\,A_\theta\right) + \frac{1}{r\,\sin\theta}\frac{\partial A_\phi}{\partial \phi}. \tag{1.15}$$

Fig. 1.2 Spherical coordinates

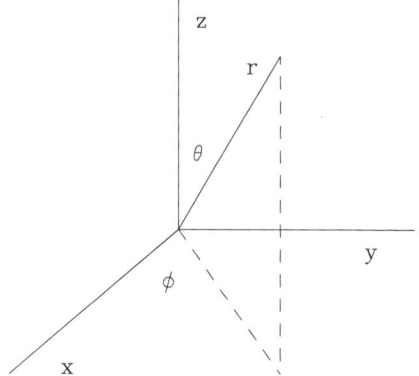

Taking the above relations into account, the continuity equation (1.6) can be written as:

$$\frac{\partial \rho}{\partial t} + \frac{1}{r^2} \frac{\partial}{\partial r} (r^2 \rho v_r) + \frac{1}{r \sin \theta} \frac{\partial}{\partial \theta} (\sin \theta \, \rho \, v_\theta) + \frac{1}{r \sin \theta} \frac{\partial}{\partial \phi} (\rho \, v_\phi) = 0, \quad (1.16)$$

where (v_r, v_θ, v_ϕ) are the coordinates of vector $\mathbf{v}$. With spherical symmetry, $v_\theta = v_\phi = 0$, the flow is unidimensional and the continuity equation is

$$\frac{\partial \rho}{\partial t} + \frac{1}{r^2} \frac{\partial}{\partial r} (r^2 \rho v) = 0, \quad (1.17)$$

where we have taken $v_r = v$. Finally, if the flow is both unidimensional and in steady state, we have

$$\frac{1}{r^2} \frac{d}{dr} (r^2 \rho v) = 0. \quad (1.18)$$

This equation can be easily integrated, so that we have

$$r^2 \rho v = \text{constant}, \quad (1.19)$$

or

$$j \propto r^{-2}, \quad (1.20)$$

that is, the mass flux decreases with the distance squared. Finally, if the gas expands at a constant velocity v, we have

$$\rho \propto r^{-2}, \quad (1.21)$$

so that the gas density falls off with the distance squared. In the study of stellar winds, Eqs. (1.19)–(1.21) are frequently used as simple approximations.

1.6   Cylindrical Coordinates

Disks around stars are observed in many situations in stellar physics, as in the case of young stellar objects of T Tauri type, or in hot stars with Be spectral types. In these cases, the mass ejection is better described by a cylindrical coordinate system (R, ϕ, z), as shown in Fig. 1.3. The same geometry can also be used in mass transfer problems in galactic scale, as in the infall processes or mass loss by the galactic disk in studies of the chemical evolution of the Galaxy.

We have the relations

$$
\begin{cases}
x = R \cos \phi \\
y = R \sin \phi \\
z = z.
\end{cases}
\tag{1.22}
$$

The volume element dV is now:

$$
dV = R \, dR \, d\phi \, dz.
\tag{1.23}
$$

For a vector $\mathbf{A}$ with coordinates (A_R, A_ϕ, A_z), we have

$$
\nabla \cdot \mathbf{A} = \frac{1}{R} \frac{\partial (R \, A_R)}{\partial R} + \frac{1}{R} \frac{\partial A_\phi}{\partial \phi} + \frac{\partial A_z}{\partial z}.
\tag{1.24}
$$

Considering the velocity vector $\mathbf{v}$ with components (v_R, v_ϕ, v_z), we have from (1.6)

$$
\frac{\partial \rho}{\partial t} + \frac{1}{R} \frac{\partial (R \, \rho \, v_R)}{\partial R} + \frac{1}{R} \frac{\partial (\rho \, v_\phi)}{\partial \phi} + \frac{\partial (\rho \, v_z)}{\partial z} = 0,
\tag{1.25}
$$

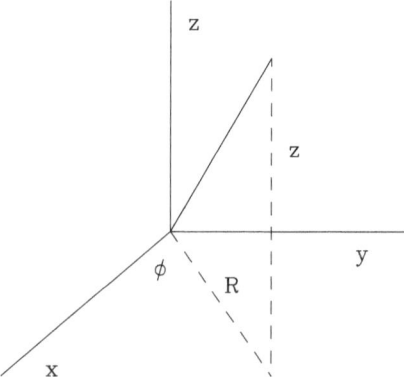

Fig. 1.3 Cylindrical coordinates

which is the continuity equation in cylindrical coordinates. With azimuthal symmetry, the equation does not depend on the azymuthal angle ϕ, and the continuity equation can be written as

$$\frac{\partial \rho}{\partial t} + \frac{1}{R} \frac{\partial (R \rho \, v_R)}{\partial R} + \frac{\partial (\rho \, v_z)}{\partial z} = 0. \tag{1.26}$$

Assuming steady state, $\partial \rho / \partial t = 0$, and the continuity equation is simplified to

$$\frac{1}{R} \frac{\partial (R \rho \, v_R)}{\partial R} + \frac{\partial (\rho \, v_z)}{\partial z} = 0. \tag{1.27}$$

1.7   The Mass Loss Rate

Equation (1.19) shows that in steady state the object is continuously supplying mass to the gas flow. Let $dM/dt = \dot{M}$ be the *mass loss rate*, which is applicable to an expanding envelope, and measured in g/s or in $M_\odot$/year. We use here the solar mass, $M_\odot = 1.99 \times 10^{33}$ g, a convenient unit in the study of stellar winds. By mass conservation, this rate must be equal to the mass per unit time that traverses a spherical layer of width dr, mass dM, located at position r, where the gas density is ρ and the velocity is $v = dr/dt$ (Fig. 1.4). We have then

$$dM = 4 \pi r^2 \rho \, dr \tag{1.28}$$

and

$$\frac{dM}{dt} = \dot{M} = 4 \pi r^2 \rho \frac{dr}{dt} = 4 \pi r^2 \rho \, v. \tag{1.29}$$

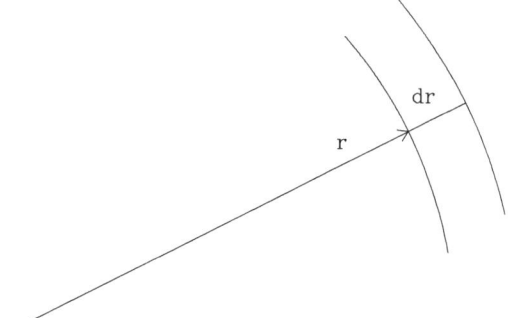

Fig. 1.4 Spherical layer with thickness dr at a distance r from the center of the star

Fig. 1.5 The solar corona, observed during the November 1944 eclipse at Chapecó, South Brazil. The hot gas expand towards the interplanetary medium, constituting the solar wind (Rodrigo Campos, LNA)

This relation defines the constant that appears in Eq. (1.19). We usually take r in cm and ρ in g/cm³. Considering that 1 year $= 3.16 \times 10^7$ s, we see that 1 $M_\odot$/year $= 6.30 \times 10^{25}$ g/s, so that we can write the relations:

$$
\begin{cases}
\dfrac{dM}{dt}\,(M_\odot/\text{year}) \simeq 1.6 \times 10^{-26}\,\dfrac{dM}{dt}\,(\text{g/s}) \\[2ex]
\dfrac{dM}{dt}\,(M_\odot/\text{year}) \simeq 2.0 \times 10^{-25}\,r^2\,\rho\,v \qquad (v \text{ in cm/s}) \\[2ex]
\dfrac{dM}{dt}\,(M_\odot/\text{year}) \simeq 2.0 \times 10^{-20}\,r^2\,\rho\,v \qquad (v \text{ in km/s}).
\end{cases}
\tag{1.30}
$$

1.7.1 Example: The Solar Wind

The Sun loses mass to the interplanetary medium, as observed by the "solar wind", a flow of charged particles that reaches the Earth orbit and beyond. We can use Eq. (1.30) and estimate the solar mass loss rate and, possibly, of other dwarf stars. The observed proton density at a distance of one astronomical unit, 1 AU $= 1.5 \times 10^{13}$ cm from the Sun is $n_p \sim 10\,\text{cm}^{-3}$, with velocity $v \sim 400$ km/s. From (1.11), the mass flow is $j \sim \rho v \sim n_p m_H v \sim 6.7 \times 10^{-16}$ g cm^{-2} s^{-1}, and the particle flux is $j' \sim n_p v \sim 4.0 \times 10^8$ cm^{-2} s^{-1}. From (1.30), the solar mass loss rate is given by $dM/dt \sim 3.0 \times 10^{-14}$ $M_\odot$/year. We know that the Sun and other stars have a chromosphere and a hot extended corona, which are heated by non-thermal processes. The corona reaches temperatures of the order of one million degrees kelvin, much higher than the average temperature of the interplanetary medium, and its evaporation basically constitutes the solar wind (Fig. 1.5). Assuming that the mass loss rate has remained constant during the solar lifetime, which is about $t_\odot \sim 4.5 \times 10^9$ year, we can estimate the total mass lost by the sun as

$$
M \sim (dM/dt)\,t_\odot \sim (3.0 \times 10^{-14})\,(4.5 \times 10^9) \simeq 1.35 \times 10^{-4}\,M_\odot,
$$

which is much smaller than the sun's present mass. The mass loss rate can be compared with the solar thermonuclear mass loss rate, that is, the rate needed to keep the present solar luminosity, $L_\odot = 3.85 \times 10^{33}$ erg/s. The thermonuclear rate is given by $(dM/dt)_T \sim L_\odot/c^2 \sim 6.8 \times 10^{-14} \, M_\odot$/year, where $c = 3.00 \times 10^{10}$ cm/s is the speed of light in vacuum, so that the thermonuclear rate is about a factor of 2 higher than the mass loss rate.

1.7.2   Example: Mass loss in Red Giants

Red giant stars are cool and luminous, and present evidences of slow winds, with typical velocities of the order of $v \sim 10$ km/s. The dimensions of the envelopes are of the order of the solar system, or $r \sim 10^{14}$ cm, and the number densities of gas particles are of the order of $n \sim 3 \times 10^8 \, cm^{-3}$, corresponding to a mass density of $\rho \sim n \, m_H \sim 5 \times 10^{-16}$ g/cm^3. Adopting these values, we obtain from (1.30) the mass loss rate $dM/dt \sim 10^{-6} \, M_\odot$/year, which is typical for a red giant star. Several physical processes have been considered as responsible for the mass ejection, involving especially the stellar radiation pressure on grains and molecules, pulsation, and wave propagation in the stellar envelopes.

1.7.3   Example: Mass Loss in Hot Stars

Hot stars, such as B-type supergiants, display fast winds, with typical velocities of the order of $v \sim 2,000$ km/s. These stars have luminosities of the order of $L \sim 10^5 \, L_\odot$, where $L_\odot \simeq 3.85 \times 10^{33}$ erg/s is the solar luminosity. Considering a typical effective temperature $T_{eff} \simeq 20,000$ K for these stars, taken as spherical, we can estimate their radius by the relation

$$L = 4 \pi R^2 \sigma T_{eff}^4, \tag{1.31}$$

which corresponds essentially to the definition of the effective temperature, where $\sigma = 5.67 \times 10^{-5}$ erg cm^{-2} s^{-1} K^{-4} is the Stefan-Boltzmann constant. We obtain $R \simeq 1.8 \times 10^{12}$ cm $\simeq 26 \, R_\odot$, where $R_\odot \simeq 6.96 \times 10^{10}$ cm is the solar radius. Assuming that the winds originate near the stellar surface and are rapidly accelerated, we can take $r \sim 2 \, R$. Considering an average density in this region of the order of $\rho \sim 10^{-14}$ g/cm^3, we estimate the mass loss rate as $dM/dt \sim 5 \times 10^{-6} \, M_\odot$/year. In these stars, radiation pressure certainly plays a fundamental role in the mass ejection, particularly through momentum absorption of the photons from the radiation field by ions located in the external layers of the stars.

Exercises

1.1. Show that the continuum hypothesis is valid in the air contained in a typical classroom.

1.2. Let μ be the density of any physical quantity, which may be the electric charge, mass, energy, etc. Calling Q the amount of this quantity produced by cubic centimeter per second, show that the conservation equation for μ can be written as

$$\frac{\partial \mu}{\partial t} + \nabla \cdot (\mu \, \mathbf{v}) = Q.$$

1.3. Prove relations (1.30).

1.4. The mass flux $\mathbf{j} = \rho \, \mathbf{v}$ (units: $g\,cm^{-2}\,s^{-1}$) can also be considered as the momentum density, or momentum per unit volume (units: $g\,cm\,s^{-1}\,cm^{-3}$). Considering the Sun has radius $R_\odot = 6.96 \times 10^{10}\,cm$, determine the mass flux near the solar surface according to the data in Example 1.7.1 Compare your result with the value obtained at $r = 1\,AU$.

1.5. The mass loss rate of the central star of the planetary nebula He2-99, measured from the analysis of the chemical composition of the wind, is $\dot{M} \simeq 4 \times 10^{-6}\,M_\odot/year$. The wind reaches a terminal velocity of $1{,}200\,km/s$. The stellar luminosity is $\log(L/L_\odot) \simeq 3.2$ and its surface temperature is $27{,}000\,K$. (a) Assuming that this velocity is reached at $r \simeq 10\,R$, where R is the stellar radius, what is the mass density of the wind in this region? (b) Assume that the wind is made only of H and He, in a proportion of 10 H atoms for each He atom. What is the average particle density in the wind?

Bibliography

Batchelor, G.K.: An Introduction to Fluid Dynamics. Cambridge University Press, Cambridge (1967) [Recent edition: 2000] (Originally published in 1967, this is one of the main basic texts on hydrodynamics in an advanced level)

Battaner, E.: Astrophysical Fluid Dynamics. Cambridge University Pess, Cambridge (1996) (More recent book with applications of fluid mechanics to the main astrophysical problems)

Choudhuri, A.R.: The Physics of Fluids and Plasmas. Cambridge University Press, Cambridge (1998) (An excellent introduction to the physics of fluids and plasmas, with several astrophysical applications. Includes a good discussion of perfect and viscous neutral fluids, gas dynamics and plasma theory)

Clarke, C., Carswell, B.: Principles of Astrophysical Fluid Dynamics. Cambridge University Press, Cambridge (2007) (A comprehensive textbook on fluid dynamics as applied to astrophysical phenomena. Discusses the main hydrodynamics equations with applications to shock waves, stellar winds, and accretion disks)

Lamers, H.J.G.L.M., Cassinelli, J.P.: Introduction to Stellar Winds. Cambridge University Press, Cambridge (1999) (The basic book on the study of stellar winds, including an advanced discussion of all the main aspects of mass loss processes, both from the observational point of view and theoretical models. Assumes an elementary knowledge of fluid mechanics)

Landau, L., Lifchitz, E.: Mécanique des Fluides. MIR, Moscou (1971) (English edition: Fluid Mechanics. Butterworth-Heinemann, Boston (1987). Excellent basic book on fluid mechanics, presenting the hydrodynamics equations and classical applications)

Maciel, W.J.: Introdução à Estrutura e Evolução Estelar. Edusp, São Paulo (1999) (Introductory book on stellar physics, discussing the main physical processes in stellar interiors and presenting order of magnitude estimates of the main physical quantities in stars)

Mihalas, D.: Stellar Atmospheres. Freeman, San Francisco (1978) (Classic advanced book on stellar atmospheres, including a chapter on stellar mass loss, stellar winds and radiative transfer. A new edition of this book is expected to be released in 2015)

Pert, G.J.: Introductory Fluid Mechanics for Physicists and Mathematicians. Wiley, New York (2013) (Recent basic introductory text on fluid mechanics especially directed to physicists and mathematicians)

Thompson, M.J.: An Introduction to Astrophysical Fuid Dynamics. Imperial College Press, London (2006) (An in-depth introduction to astrophysical fluid dynamics, emphasizing observational phenomena and simple models of astrophysical flows)

Trefil, J.S.: Introduction to the Physics of Fluids and Solids. Dover, New York (2010) (Paperback edition of a 1975 book. A good introduction to the mechanics of fluids and solids, with some applications to astrophysical problems)

Chapter 2
The Euler Equation

Abstract This chapter presents Euler equation without and with the presence of a magnetic field. The equation is written in both spherical and cylindrical coordinates, and applied to a stellar envelope. Some examples involving the solar wind are given, and a discussion is presented on the escape velocity from stars. The chapter ends with some further examples applied to hot and cool stars.

2.1 The Euler Equation

A fluid in equilibrium is not affected by any shearing stress, that is, there are no forces acting in the horizontal direction (Fig. 2.1a). If the fluid is in motion, these forces will be important, as a result of the *viscosity* of the fluid. Let us consider initially an *ideal fluid*, for which the viscosity is negligible. Therefore, the horizontal stress forces that arise between the moving fluid layers are negligible (Fig. 2.1b). In this case, apart from external forces, the only stress that has to be considered in the fluid is the *normal* stress, or pressure P. Pressure is a scalar quantity, and a function of the space coordinates and time, $P(x, y, z, t)$.

In the simplest cases, the fluid particles move in layers, or sheets, which constitute a *laminar* flow. If the particle trajectories are irregular, the flow is *turbulent*. In this book we will consider essentially turbulent free flows. In the general case, the displacements of a volume element in the fluid include its *translation*, *rotation*, and *deformation*. In this book, we will be interested principally in the translation motion, that is, we will consider non rotating fluids, in which the angular velocity of a fluid volume element is $\omega = 0$, and non-viscous fluids, in which the deformation (contraction or stretching) is negligible.

Let us consider again a volume V in the fluid. The total force acting on this volume due to the interactions with the remaining fluid particles is

$$- \oint P \, \mathbf{n} \, dS$$

W.J. Maciel, *Hydrodynamics and Stellar Winds: An Introduction*, Undergraduate Lecture Notes in Physics, DOI 10.1007/978-3-319-04328-9_2,
© Springer International Publishing Switzerland 2014

Fig. 2.1 (**a**) Fluid in
equilibrium. (**b**) Fluid moving
under the action of viscous
forces

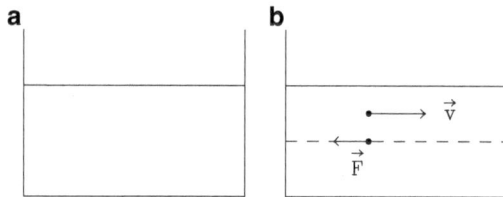

(see Fig. 1.1), where the negative sign (−) is due to the fact that the force acts *on* the considered element. Applying the "gradient theorem" to a scalar quantity A, we have

$$\oint A \mathbf{n} \, dS = \int_V \nabla A \, dV , \tag{2.1}$$

so that

$$-\oint P \mathbf{n} \, dS = -\int_V \nabla P \, dV . \tag{2.2}$$

Therefore, the force by the remaining parts of the fluid on the volume element dV is $-\nabla P \, dV$, and the force per unit volume is simply $-\nabla P$. We can then express the momentum conservation by

$$\rho \times \text{acceleration} = -\nabla P \tag{2.3}$$

where ρ is again the fluid density. In the case of a particle of mass m under the action of a force $\mathbf{F}$ and moving with velocity $\mathbf{v}$, the acceleration is simply $\mathbf{a} = \mathbf{F}/m = d\mathbf{v}/dt$. For a "fluid particle", the velocity variation has two components:

$$d\mathbf{v}_1 = \frac{\partial \mathbf{v}}{\partial t} \, dt , \tag{2.4}$$

which refers to the velocity variation at a *fixed point in space*, that is, having constant coordinates x, y, z in a time interval dt, and

$$d\mathbf{v}_2 = dx \, \frac{\partial \mathbf{v}}{\partial x} + dy \, \frac{\partial \mathbf{v}}{\partial y} + dz \, \frac{\partial \mathbf{v}}{\partial z} , \tag{2.5}$$

which refers to the velocity variation *at a given time* between two different points in space separated by

$$d\mathbf{r} = dx \, \mathbf{i} + dy \, \mathbf{j} + dz \, \mathbf{k} , \tag{2.6}$$

which is the distance covered by the fluid particle in time dt. ($\mathbf{i}, \mathbf{j}, \mathbf{k}$ are the unit vectors of the directions x, y, z). Recalling that

$$\nabla = \frac{\partial}{\partial x}\,\mathbf{i} + \frac{\partial}{\partial y}\,\mathbf{j} + \frac{\partial}{\partial z}\,\mathbf{k}\,, \tag{2.7}$$

we have

$$d\mathbf{r} \cdot \nabla = dx\,\frac{\partial}{\partial x} + dy\,\frac{\partial}{\partial y} + dz\,\frac{\partial}{\partial z}\,, \tag{2.8}$$

so that

$$d\mathbf{v}_2 = (d\mathbf{r} \cdot \nabla)\,\mathbf{v}\,. \tag{2.9}$$

The total variation $d\mathbf{v}$ is

$$d\mathbf{v} = d\mathbf{v}_1 + d\mathbf{v}_2 = \frac{\partial \mathbf{v}}{\partial t}\,dt + (d\mathbf{r} \cdot \nabla)\,\mathbf{v}\,. \tag{2.10}$$

The acceleration is then

$$\text{acceleration} = \frac{\partial \mathbf{v}}{\partial t} + (\mathbf{v} \cdot \nabla)\,\mathbf{v}\,. \tag{2.11}$$

The derivative in (2.11) is the *total derivative*, which is usually represented as

$$\frac{D}{Dt} = \frac{\partial}{\partial t} + (\mathbf{v} \cdot \nabla)\,. \tag{2.12}$$

The description of the fluid motion considering separately two terms, that is, $\partial/\partial t$ in a fixed position and $\mathbf{v} \cdot \nabla$ in a given time, corresponds to the *Eulerian description*. On the other hand, the description of the fluid motion in terms of the total derivative D/Dt corresponds to the *Lagrangian description*. In this case, we can imagine that we are following the motion of a fluid element. Naturally, the Lagrangian derivative D/Dt is related to the Eulerian derivatives by Eq. (2.12). Considering (2.11) and (2.3), the equation of motion is

$$\begin{cases} \rho\left[\dfrac{\partial \mathbf{v}}{\partial t} + (\mathbf{v} \cdot \nabla)\,\mathbf{v}\right] = -\nabla P \\[2mm] \dfrac{\partial \mathbf{v}}{\partial t} + (\mathbf{v} \cdot \nabla)\,\mathbf{v} = -\dfrac{1}{\rho}\,\nabla P \\[2mm] \dfrac{D\mathbf{v}}{Dt} = -\dfrac{1}{\rho}\,\nabla P\,. \end{cases} \tag{2.13}$$

Equations (2.13) are different forms of the *Euler equation*, which was originally derived in 1755.

An important application of the equation of motion occurs when the fluid is in a gravitational field characterized by the acceleration (force per unit mass) $\mathbf{g}$. In this case, each volume unit is under the action of the force $\rho\,\mathbf{g}$, which is the gravitational force acting per unit volume, so that the equation of motion (2.13) is written as

$$\frac{\partial \mathbf{v}}{\partial t} + (\mathbf{v} \cdot \nabla)\,\mathbf{v} = -\frac{1}{\rho}\,\nabla P + \mathbf{g}\,. \tag{2.14}$$

More generally, if the fluid is under the action of an external force field $\mathbf{F}$ (dyn/cm^3), that is, $\mathbf{F}$ is the force acting on a unit volume, we have

$$\frac{\partial \mathbf{v}}{\partial t} + (\mathbf{v} \cdot \nabla)\,\mathbf{v} = -\frac{1}{\rho}\,\nabla P + \frac{1}{\rho}\,\mathbf{F}\,. \tag{2.15}$$

2.1.1 Example: Static Star

Let us analyze a simple application of Eq. (2.14), considering a static star, that is, taking $\mathbf{v} = 0$. In this case, the pressure forces exactly balance the gravitational force in each fluid element constituting the star. We can then assume that the gravitational force can be derived from a gravitational potential such that $\mathbf{g} = -\nabla\phi$. The potential ϕ can be related to the gas density in the star by Poisson equation,

$$\nabla^2 \phi = 4\,\pi\,G\,\rho\,. \tag{2.16}$$

From (2.14) with $\mathbf{v} = 0$, we have

$$\frac{1}{\rho}\,\nabla P = \mathbf{g} = -\nabla\phi\,.$$

Taking the divergence of this equation,

$$\nabla \cdot \left(\frac{1}{\rho}\,\nabla P\right) = -\nabla \cdot (\nabla\phi) = -\nabla^2 \phi\,,$$

that is,

$$\nabla \cdot \left(\frac{1}{\rho}\,\nabla P\right) = -4\,\pi\,G\,\rho\,. \tag{2.17}$$

In order to solve (2.17) we need an independent relation involving P and ρ, as for example the equation of state. In this book, however, we will be especially interested in expanding stellar envelopes, so that the more appropriate form of Euler equation is (2.14) or (2.15).

2.2   Euler Equation in the Presence of a Magnetic Field

We saw in Chap. 1 that the continuum hypothesis can be applied to several astrophysical situations, particularly if the particle mean free path, which is essentially established by collisions, is small relative to some typical dimension of the system. We can then say that collisions favour the behaviour of a system as a continuous fluid. Magnetic fields in a plasma are also able to keep charged particles locally confined for a sufficiently long time so that the system behaves as a fluid, even if the collisional processes are not efficient.

A detailed treatment of the astrophysical applications of magnetohydrodynamics (MHD) is beyond the limits of this book, and the interested reader may check some of the recent texts listed at the end of Chap. 1, such as Choudhuri (1998). In this section, we will limit ourselves to mentioning the main changes in Eq. (2.15) due to presence of a magnetic field. In this case, apart from the two thermodynamic quantities (usually the pressure P and density ρ) and the three velocity components, it is necessary to especify the vector magnetic field $\mathbf{B}$ so that the plasma can be described in a magnetohydrodynamic model. This type of model is frequently useful in the astrophysical applications, in particular in non-relativistic processes in which the plasma motions are essentially constant, or vary slowly with time, that is, when the time scale of the physical processes τ is much larger than the inverse of the plasma frequency, ω_p, or $\tau \gg 1/\omega_p$. These conditions are generally applied to the stellar winds, in the cases where the plasma motions are due to the action of mechanic and magnetic forces.

The continuity equation (1.6) remains the same, since the hypothesis of no sources or sinks is still valid. For the Euler equation (2.15) we may write

$$\frac{\partial \mathbf{v}}{\partial t} + (\mathbf{v} \cdot \nabla)\, \mathbf{v} = -\frac{1}{\rho}\, \nabla P + \frac{1}{\rho}\, \mathbf{F} + \frac{1}{\rho c}\, \mathbf{J} \times \mathbf{B}\,, \qquad (2.18)$$

where $\mathbf{B}$ is the magnetic field and $\mathbf{J}$ is electric current density. As in the remaining applications in this book, we use the cgs system of units, in which the electric quantities are in e.s.u. and the magnetic quantities in e.m.u.

A detailed derivation of Eq. (2.18) from basic microscopic principles can be found in Choudhuri (1998, Chap. 13). In the MHD model we can write Maxwell equation without the displacement current term as

$$\nabla \times \mathbf{B} = \frac{4\pi}{c}\, \mathbf{J}\,, \qquad (2.19)$$

that is, the current density $\mathbf{J}$ can be obtained from the magnetic field $\mathbf{B}$ and vice-versa. Substituting (2.19) in (2.18) we have

$$\frac{\partial \mathbf{v}}{\partial t} + (\mathbf{v} \cdot \nabla)\, \mathbf{v} = -\frac{1}{\rho}\, \nabla P + \frac{1}{\rho}\, \mathbf{F} + \frac{1}{4\pi\rho}\, (\nabla \times \mathbf{B}) \times \mathbf{B} \qquad (2.20)$$

Using the vector identity

$$\nabla(\mathbf{A} \cdot \mathbf{B}) = \mathbf{A} \times (\nabla \times \mathbf{B}) + \mathbf{B} \times (\nabla \times \mathbf{A}) + (\mathbf{A} \cdot \nabla)\mathbf{B} + (\mathbf{B} \cdot \nabla)\mathbf{A}$$

and taking $\mathbf{A} = \mathbf{B}$, it is easy to show that

$$(\nabla \times \mathbf{B}) \times \mathbf{B} = (\mathbf{B} \cdot \nabla)\mathbf{B} - \nabla\left(\frac{B^2}{2}\right).$$

Substituting this equation in (2.20) we obtain the alternative form of Euler equation,

$$\frac{\partial \mathbf{v}}{\partial t} + (\mathbf{v} \cdot \nabla)\mathbf{v} = \frac{1}{\rho}\,\mathbf{F} - \frac{1}{\rho}\,\nabla\left(P + \frac{B^2}{8\pi}\right) + \frac{(\mathbf{B} \cdot \nabla)\mathbf{B}}{4\pi\rho}\,. \qquad (2.21)$$

From Eq. (2.21) we see that the term $B^2/8\pi$ corresponds to the magnetic pressure, introduced by the presence of the magnetic field. The last term in (2.21) is related to the tension along the magnetic field lines, that is, the magnetic field is associated to an isotropic pressure and also to a tension along the field lines.

2.2.1   Example: Solar Atmosphere

As an example, we may compare the magnetic pressure $B^2/8\pi$ with the gas pressure in a few points in the solar atmosphere. In the photosphere, the pressure varies by several orders of magnitude, but we can consider an average point where the total particle density is of the order of 10^{17} cm^{-3} and the temperature is about 6,000 K, so that the total pressure is 10^5 dyn/cm^2. Similar pressures are obtained by a magnetic field of the order of 1,500 Gauss. These values are observed in regions of the solar atmosphere such the plages and solar faculae, and still higher values, $B \sim 3{,}000$ Gauss, are associated to the umbral regions of the sunspots. Here the temperature tends to be somewhat lower than in the photosphere, but the higher densities lead to pressures of the order of 10^5 dyn/cm^2 or higher. In the outer regions, which are associated to the solar wind, the field is probably lower by a few orders of magnitude, reaching values near the earth magnetic field, $B \sim 1$ Gauss. In this case, the magnetic pressure is of the order of 10^{-2} dyn/cm^2, which can be compared with the gas pressure in a coronal region about 2 solar radii from the solar surface, where the density $n \sim 10^6$ cm^{-3}, $T \sim 10^6$ K and the pressure is $P \sim 10^{-3}$ dyn/cm^2 (see Example 2.5.2).

Magnetic fields are also associated to the atmospheres and envelopes of red giant stars, although their magnitudes are uncertain. For example, in the envelope of a red giant with $T \sim 10^3$ K and $n \sim 10^8$ cm^{-3}, the gas pressure is of the order of 10^{-5} dyn/cm^2, which is equivalent to the magnetic pressure if the field is just $B \sim 10^{-2}$ Gauss, suggesting that magnetic fields effectively have some influence on the dynamics of the circumstellar envelopes.

2.3 Spherical Coordinates

Let us derive the equation of motion in the form (2.15) in a system of spherical coordinates (r, θ, ϕ), defined in Fig. 2.2. The unit vectors $(\mathbf{n}, \boldsymbol{\ell}, \mathbf{m})$ are related to the unit vectors $(\mathbf{i}, \mathbf{j}, \mathbf{k})$ corresponding to the directions x, y, z by

$$\begin{cases} \mathbf{n} = \sin\theta \, \cos\phi \, \mathbf{i} + \sin\theta \, \sin\phi \, \mathbf{j} + \cos\theta \, \mathbf{k} \\ \boldsymbol{\ell} = \cos\theta \, \cos\phi \, \mathbf{i} + \cos\theta \, \sin\phi \, \mathbf{j} - \sin\theta \, \mathbf{k} \\ \mathbf{m} = -\sin\phi \, \mathbf{i} + \cos\phi \, \mathbf{j} \, . \end{cases} \tag{2.22}$$

The position vector $\mathbf{r}$ and velocity $\mathbf{v}$ are given by

$$\mathbf{r} = r \, \mathbf{n}(\theta, \phi) \tag{2.23}$$

$$\mathbf{v} = \dot{\mathbf{r}} = \dot{r} \, \mathbf{n} + r \, \dot{\mathbf{n}} \, . \tag{2.24}$$

On the other hand, from (2.22) we have the relations

$$\begin{cases} \dfrac{\partial \mathbf{n}}{\partial \theta} = \boldsymbol{\ell} \\[2mm] \dfrac{\partial \mathbf{n}}{\partial \phi} = \sin\theta \, \mathbf{m} \, , \end{cases} \tag{2.25}$$

so that

$$\dot{\mathbf{n}} = \frac{\partial \mathbf{n}}{\partial \theta} \, \dot{\theta} + \frac{\partial \mathbf{n}}{\partial \phi} \, \dot{\phi} = \dot{\theta} \, \boldsymbol{\ell} + \dot{\phi} \, \sin\theta \, \mathbf{m} \, , \tag{2.26}$$

and therefore

$$\mathbf{v} = \dot{r} \, \mathbf{n} + r(\dot{\theta} \, \boldsymbol{\ell} + \dot{\phi} \, \sin\theta \, \mathbf{m}) \, . \tag{2.27}$$

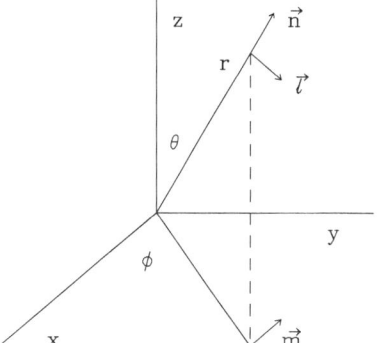

Fig. 2.2 Spherical coordinates and unit vectors in a cartesian axis system

We may write

$$\mathbf{v} = v_r \, \mathbf{n} + v_\theta \, \boldsymbol{\ell} + v_\phi \, \mathbf{m} \, , \tag{2.28}$$

where

$$\begin{cases} v_r = \dot{r} \\ v_\theta = r \, \dot{\theta} \\ v_\phi = r \, \dot{\phi} \, \sin \theta \, . \end{cases} \tag{2.29}$$

We have also

$$\begin{cases} \dot{v}_r = \ddot{r} \\ \dot{v}_\theta = \dot{r} \, \dot{\theta} + r \, \ddot{\theta} \\ \dot{v}_\phi = \dot{r} \, \dot{\phi} \, \sin \theta + r \, \ddot{\phi} \, \sin \theta + r \, \dot{\phi} \, \cos \theta \, \dot{\theta} \, . \end{cases} \tag{2.30}$$

Still from (2.22), we have

$$\begin{cases} \dfrac{\partial \boldsymbol{\ell}}{\partial \theta} = -\mathbf{n} \\ \dfrac{\partial \boldsymbol{\ell}}{\partial \phi} = \cos \theta \, \mathbf{m} \, , \end{cases} \tag{2.31}$$

so that

$$\dot{\boldsymbol{\ell}} = \frac{\partial \boldsymbol{\ell}}{\partial \theta} \, \dot{\theta} + \frac{\partial \boldsymbol{\ell}}{\partial \phi} \, \dot{\phi} = -\dot{\theta} \, \mathbf{n} + \dot{\phi} \, \cos \theta \, \mathbf{m} \, . \tag{2.32}$$

From (2.22), we may also write

$$\begin{cases} \dfrac{\partial \mathbf{m}}{\partial \theta} = 0 \\ \dfrac{\partial \mathbf{m}}{\partial \phi} = -\sin \theta \, \mathbf{n} - \cos \theta \, \boldsymbol{\ell} \, , \end{cases} \tag{2.33}$$

so that

$$\dot{\mathbf{m}} = \frac{\partial \mathbf{m}}{\partial \theta} \, \dot{\theta} + \frac{\partial \mathbf{m}}{\partial \phi} \, \dot{\phi} = -\dot{\phi} \, \sin \theta \, \mathbf{n} - \dot{\phi} \, \cos \theta \, \boldsymbol{\ell} \, . \tag{2.34}$$

Using (2.26), (2.28), (2.32) and (2.34), we have

$$\dot{\mathbf{v}} = \ddot{\mathbf{r}} = \dot{v}_r \, \mathbf{n} + v_r \, \dot{\mathbf{n}} + \dot{v}_\theta \, \boldsymbol{\ell} + v_\theta \, \dot{\boldsymbol{\ell}} + \dot{v}_\phi \, \mathbf{m} + v_\phi \, \dot{\mathbf{m}}$$

$$= \dot{v}_r \, \mathbf{n} + v_r \, (\dot{\theta} \, \boldsymbol{\ell} + \dot{\phi} \, \sin \theta \, \mathbf{m}) + \dot{v}_\theta \, \boldsymbol{\ell} + v_\theta (-\dot{\theta} \, \mathbf{n} + \dot{\phi} \, \cos \theta \, \mathbf{m})$$

$$+ \dot{v}_\phi \, \mathbf{m} + v_\phi (-\dot{\phi} \, \sin \theta \, \mathbf{n} - \dot{\phi} \, \cos \theta \, \boldsymbol{\ell}) \, ,$$

or

$$\dot{\mathbf{v}} = (\dot{v}_r - v_\theta\,\dot\theta - v_\phi\,\dot\phi\,\sin\theta)\,\mathbf{n} + (v_r\,\dot\theta + \dot{v}_\theta - v_\phi\,\dot\phi\,\cos\theta)\,\boldsymbol{\ell}$$

$$+ (v_r\,\dot\phi\,\sin\theta + v_\theta\,\dot\phi\,\cos\theta + \dot{v}_\phi)\,\mathbf{m}\ .$$

Using Eq. (2.29)

$$\dot{\mathbf{v}} = \left(\dot{v}_r - \frac{v_\theta^2}{r} - \frac{v_\phi^2}{r}\right)\mathbf{n} + \left(\frac{v_r\,v_\theta}{r} + \dot{v}_\theta - \frac{v_\phi^2\cos\theta}{r\sin\theta}\right)\boldsymbol{\ell} + \left(\frac{v_r\,v_\phi}{r} + \frac{v_\theta\,v_\phi\cos\theta}{r\sin\theta} + \dot{v}_\phi\right)\mathbf{m}\ ,$$

$$(2.35)$$

or still, using (2.29) and (2.30),

$$\dot{\mathbf{v}} = (\ddot{r} - r\,\dot\theta^2 - r\,\dot\phi^2\sin^2\theta)\,\mathbf{n} + (r\,\ddot\theta + 2\,\dot{r}\,\dot\theta - r\,\dot\phi^2\sin\theta\cos\theta)\,\boldsymbol{\ell}$$

$$+ (r\,\ddot\phi\,\sin\theta + 2\,\dot{r}\,\dot\phi\,\sin\theta + 2\,r\,\dot\theta\,\dot\phi\,\cos\theta)\,\mathbf{m}\ . \qquad (2.36)$$

Using (2.28) and the operator ∇ in spherical coordinates:

$$\nabla = \frac{\partial}{\partial r}\,\mathbf{n} + \frac{1}{r}\frac{\partial}{\partial\theta}\,\boldsymbol{\ell} + \frac{1}{r\sin\theta}\frac{\partial}{\partial\phi}\,\mathbf{m}\ , \qquad (2.37)$$

$$\mathbf{v}\cdot\nabla = v_r\frac{\partial}{\partial r} + \frac{v_\theta}{r}\frac{\partial}{\partial\theta} + \frac{v_\phi}{r\sin\theta}\frac{\partial}{\partial\phi}\ . \qquad (2.38)$$

Applying this operator to vector $\mathbf{v}$,

$$(\mathbf{v}\cdot\nabla)\mathbf{v} = \left(v_r\frac{\partial v_r}{\partial r}\mathbf{n} + v_r\frac{\partial v_\theta}{\partial r}\boldsymbol{\ell} + v_r\frac{\partial v_\phi}{\partial r}\mathbf{m}\right) + \left(\frac{v_\theta}{r}\frac{\partial v_r}{\partial\theta}\mathbf{n} + \right.$$

$$\frac{v_\theta}{r}\frac{\partial\mathbf{n}}{\partial\theta}v_r + \frac{v_\theta}{r}\frac{\partial v_\theta}{\partial\theta}\boldsymbol{\ell} + \frac{v_\theta^2}{r}\frac{\partial\boldsymbol{\ell}}{\partial\theta} + \frac{v_\theta}{r}\frac{\partial v_\phi}{\partial\theta}\mathbf{m} + \frac{v_\theta\,v_\phi}{r}\frac{\partial\mathbf{m}}{\partial\theta}\left.\right) +$$

$$\left(\frac{v_\phi}{r\sin\theta}\frac{\partial v_r}{\partial\phi}\mathbf{n} + \frac{v_\phi}{r\sin\theta}\frac{\partial\mathbf{n}}{\partial\phi}v_r + \frac{v_\phi}{r\sin\theta}\frac{\partial v_\theta}{\partial\phi}\boldsymbol{\ell} + \frac{v_\phi}{r\sin\theta}\frac{\partial\boldsymbol{\ell}}{\partial\phi}v_\theta + \right.$$

$$\frac{v_\phi}{r\sin\theta}\frac{\partial v_\phi}{\partial\phi}\mathbf{m} + \frac{v_\phi^2}{r\sin\theta}\frac{\partial\mathbf{m}}{\partial\phi}\left.\right)$$

Using (2.25), (2.31) and (2.33), we have

$$(\mathbf{v}\cdot\nabla)\mathbf{v} = \left(v_r\frac{\partial v_r}{\partial r} + \frac{v_\theta}{r}\frac{\partial v_r}{\partial\theta} - \frac{v_\theta^2}{r} + \frac{v_\phi}{r\sin\theta}\frac{\partial v_r}{\partial\phi} - \frac{v_\phi^2}{r}\right)\mathbf{n} +$$

$$\left(v_r\frac{\partial v_\theta}{\partial r} + \frac{v_r\,v_\theta}{r} + \frac{v_\theta}{r}\frac{\partial v_\theta}{\partial\theta} + \frac{v_\phi}{r\sin\theta}\frac{\partial v_\theta}{\partial\phi} - \frac{v_\phi^2\cos\theta}{r\,\sin\theta}\right)\boldsymbol{\ell} +$$

$$\left(v_r \, \frac{\partial v_\phi}{\partial r} + \frac{v_\theta}{r} \, \frac{\partial v_\phi}{\partial \theta} + \frac{v_r \, v_\phi}{r} + \frac{v_\theta \, v_\phi \cos \theta}{r \sin \theta} + \frac{v_\phi}{r \sin \theta} \, \frac{\partial v_\phi}{\partial \phi} \right) \mathbf{m}$$

$$(2.39)$$

or

$$(\mathbf{v} \cdot \nabla) \mathbf{v} = \left(\mathbf{v} \cdot \nabla v_r - \frac{v_\theta^2}{r} - \frac{v_\phi^2}{r} \right) \mathbf{n} + \left(\mathbf{v} \cdot \nabla v_\theta + \frac{v_r \, v_\theta}{r} - \frac{v_\phi^2 \cot \theta}{r} \right) \boldsymbol{\ell} +$$

$$\left(\mathbf{v} \cdot \nabla v_\phi + \frac{v_r \, v_\phi}{r} + \frac{v_\theta \, v_\phi \cot \theta}{r} \right) \mathbf{m} \, .$$

$$(2.40)$$

Considering (2.28), the first term of the first member of Eq. (2.15) can be written as

$$\frac{\partial \mathbf{v}}{\partial t} = \frac{\partial v_r}{\partial t} \, \mathbf{n} + \frac{\partial v_\theta}{\partial t} \, \boldsymbol{\ell} + \frac{\partial v_\phi}{\partial t} \, \mathbf{m} \, .$$

$$(2.41)$$

From (2.37), the pressure term in (2.15) can be written as

$$\nabla P = \frac{\partial P}{\partial r} \, \mathbf{n} + \frac{1}{r} \, \frac{\partial P}{\partial \theta} \, \boldsymbol{\ell} + \frac{1}{r \sin \theta} \, \frac{\partial P}{\partial \phi} \, \mathbf{m} \, .$$

$$(2.42)$$

The force per unit volume can be written as

$$\mathbf{F} = F_r \, \mathbf{n} + F_\theta \, \boldsymbol{\ell} + F_\phi \, \mathbf{m} \, .$$

$$(2.43)$$

Rearranging expressions (2.39)–(2.43), we can write the equations separately, according to components $\mathbf{n}, \boldsymbol{\ell}, \mathbf{m}$:

Radial component, unit vector $\mathbf{n}$:

$$\frac{\partial v_r}{\partial t} + v_r \, \frac{\partial v_r}{\partial r} + \frac{v_\theta}{r} \, \frac{\partial v_r}{\partial \theta} + \frac{v_\phi}{r \sin \theta} \, \frac{\partial v_r}{\partial \phi} - \frac{v_\theta^2 + v_\phi^2}{r} = -\frac{1}{\rho} \, \frac{\partial P}{\partial r} + \frac{1}{\rho} \, F_r \quad (2.44)$$

$$\frac{\partial v_r}{\partial t} + \mathbf{v} \cdot \nabla v_r - \frac{v_\theta^2 + v_\phi^2}{r} = -\frac{1}{\rho} \, \frac{\partial P}{\partial r} + \frac{1}{\rho} \, F_r \, .$$

$$(2.45)$$

Polar component, unit vector $\boldsymbol{\ell}$:

$$\frac{\partial v_\theta}{\partial t} + v_r \, \frac{\partial v_\theta}{\partial r} + \frac{v_\theta}{r} \, \frac{\partial v_\theta}{\partial \theta} + \frac{v_\phi}{r \sin \theta} \, \frac{\partial v_\theta}{\partial \phi} + \frac{v_r \, v_\theta}{r} - \frac{v_\phi^2 \cot \theta}{r} = -\frac{1}{\rho r} \, \frac{\partial P}{\partial \theta} + \frac{1}{\rho} \, F_\theta$$

$$(2.46)$$

$$\frac{\partial v_\theta}{\partial t} + \mathbf{v} \cdot \nabla v_\theta + \frac{v_r \, v_\theta}{r} - \frac{v_\phi^2 \cot \theta}{r} = -\frac{1}{\rho r} \, \frac{\partial P}{\partial \theta} + \frac{1}{\rho} \, F_\theta \, .$$

$$(2.47)$$

Azymuthal component, unit vector **m**:

$$\frac{\partial v_\phi}{\partial t} + v_r \frac{\partial v_\phi}{\partial r} + \frac{v_\theta}{r} \frac{\partial v_\phi}{\partial \theta} + \frac{v_\phi}{r \sin \theta} \frac{\partial v_\phi}{\partial \phi} + \frac{v_r \, v_\phi}{r} + \frac{v_\theta \, v_\phi \cot \theta}{r}$$

$$= -\frac{1}{\rho r \sin \theta} \frac{\partial P}{\partial \phi} + \frac{1}{\rho} F_\phi \tag{2.48}$$

$$\frac{\partial v_\phi}{\partial t} + \mathbf{v} \cdot \nabla v_\phi + \frac{v_r \, v_\phi}{r} + \frac{v_\theta \, v_\phi \cot \theta}{r} = -\frac{1}{\rho r \sin \theta} \frac{\partial P}{\partial \phi} + \frac{1}{\rho} F_\phi \, . \tag{2.49}$$

2.3.1   Example: Azymuthal Symmetry

Assuming that v_ϕ and its derivatives are zero, there are only two equations:

$$\frac{\partial v_r}{\partial t} + v_r \frac{\partial v_r}{\partial r} + \frac{v_\theta}{r} \frac{\partial v_r}{\partial \theta} - \frac{v_\theta^2}{r} = -\frac{1}{\rho} \frac{\partial P}{\partial r} + \frac{1}{\rho} F_r \tag{2.50}$$

$$\frac{\partial v_\theta}{\partial t} + v_r \frac{\partial v_\theta}{\partial r} + \frac{v_\theta}{r} \frac{\partial v_\theta}{\partial \theta} + \frac{v_r \, v_\theta}{r} = -\frac{1}{\rho r} \frac{\partial P}{\partial \theta} + \frac{1}{\rho} F_\theta \, . \tag{2.51}$$

2.3.2   Example: Spherical Symmetry

Assuming that v_θ and its derivatives are also zero, Eq. (2.50) can be written as

$$\frac{\partial v}{\partial t} + v \frac{\partial v}{\partial r} = -\frac{1}{\rho} \frac{\partial P}{\partial r} + \frac{1}{\rho} F \tag{2.52}$$

where we dropped subscript r.

2.3.3   Example: Spherical Symmetry and Steady State

In this case, $\partial v / \partial t = 0$ in (2.52), so that

$$v \frac{dv}{dr} = -\frac{1}{\rho} \frac{dP}{dr} + \frac{1}{\rho} F \, . \tag{2.53}$$

This is frequently the form of the Euler equation that is used in the study of stellar winds.

2.3.4   Example: Hydrostatic Equilibrium

A particular case of (2.53) is that of *hydrostatic equilibrium*, where $v = 0$ in all points of the fluid. In this case,

$$\frac{dP}{dr} = F \ . \tag{2.54}$$

We will consider this situation again in Sect. 2.5.

2.4   Cylindrical Coordinates

Let us concisely repeat the procedure of Sect. 2.3 in the case of a system of cylindrical coordinates (R, ϕ, z), with unit vectors $(\mathbf{h}, \mathbf{m}, \mathbf{k})$ as shown in Fig. 2.3. The relations between the cartesian coordinates (x, y, z) and (R, ϕ, z) are given by Eq. (1.22). The relations between the unit vectors $(\mathbf{h}, \mathbf{m}, \mathbf{k})$ and the corresponding vectors of the cartesian system $(\mathbf{i}, \mathbf{j}, \mathbf{k})$ are:

$$\begin{cases} \mathbf{h} = \cos \phi - + \sin \phi \, \mathbf{j} \\ \mathbf{m} = -\sin \phi \, \mathbf{i} + \cos \phi \, \mathbf{j} \\ \mathbf{k} = \mathbf{k} \, , \end{cases} \tag{2.55}$$

from which we obtain the derivatives

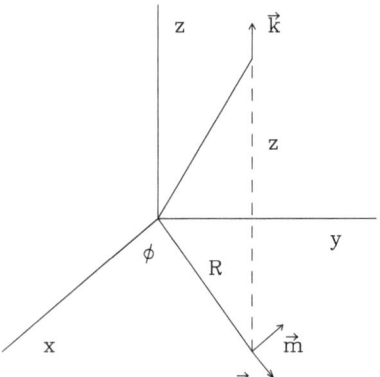

Fig. 2.3 Cylindrical coordinates and unit vectors in a cartesian axis system

$$\begin{cases} \dfrac{d\mathbf{h}}{d\phi} = \mathbf{m} \\ \dfrac{d\mathbf{m}}{d\phi} = -\mathbf{h} \ . \end{cases} \tag{2.56}$$

The position vector $\mathbf{r}$ and the velocity vector $\mathbf{v}$ are given by

$$\mathbf{r} = R\,\mathbf{h} + z\,\mathbf{k} \tag{2.57}$$

$$\mathbf{v} = \frac{d\mathbf{r}}{dt} = \dot{R}\,\mathbf{h} + R\dot{\phi}\,\mathbf{m} + \dot{z}\,\mathbf{k} \ . \tag{2.58}$$

Analogously to Eq. (2.28), the velocity can be written as

$$\mathbf{v} = v_R\,\mathbf{h} + v_\phi\,\mathbf{m} + v_z\,\mathbf{k} \ , \tag{2.59}$$

from which we see that $v_R = \dot{R}$, $v_\phi = R\dot{\phi}$ and $v_z = \dot{z}$.

The gradient in cylindrical coordinates is:

$$\nabla = \frac{\partial}{\partial R}\,\mathbf{h} + \frac{1}{R}\frac{\partial}{\partial \phi}\,\mathbf{m} + \frac{\partial}{\partial z}\,\mathbf{k} \ , \tag{2.60}$$

so that the operator $\mathbf{v} \cdot \nabla$ appearing in (2.15) is written as

$$\mathbf{v} \cdot \nabla = v_R\frac{\partial}{\partial R} + \frac{v_\phi}{R}\frac{\partial}{\partial \phi} + v_z\frac{\partial}{\partial z} \ . \tag{2.61}$$

Applying this operator to vector $\mathbf{v}$, we obtain the equivalent expression to (2.39),

$$(\mathbf{v} \cdot \nabla)\mathbf{v} = \left(v_R\frac{\partial v_R}{\partial R} + \frac{v_\phi}{R}\frac{\partial v_R}{\partial \phi} - \frac{v_\phi^2}{R} + v_z\frac{\partial v_R}{\partial z} \right)\mathbf{h} +$$

$$\left(v_R\frac{\partial v_\phi}{\partial R} + \frac{v_R v_\phi}{R} + \frac{v_\phi}{R}\frac{\partial v_\phi}{\partial \phi} + v_z\frac{\partial v_\phi}{\partial z} \right)\mathbf{m} + \left(v_R\frac{\partial v_z}{\partial R} + \frac{v_\phi}{R}\frac{\partial v_z}{\partial \phi} + v_z\frac{\partial v_z}{\partial z} \right)\mathbf{k} \ . \tag{2.62}$$

The term $\partial\mathbf{v}/\partial t$ in (2.15) can be written in terms of the cylindrical coordinates, analogously to Eq. (2.41), and the same procedure can be applied to an external force $\mathbf{F}$, with components F_R, F_ϕ, F_z. Applying again the gradient operator (2.60) to the pressure term P, and using (2.62), we can write Euler equation for the three components (R, ϕ, z) as:

Radial component, unit vector $\mathbf{h}$:

$$\frac{\partial v_R}{\partial t} + v_R\frac{\partial v_R}{\partial R} + \frac{v_\phi}{R}\frac{\partial v_R}{\partial \phi} - \frac{v_\phi^2}{R} + v_z\frac{\partial v_R}{\partial z} = -\frac{1}{\rho}\frac{\partial P}{\partial R} + \frac{1}{\rho}F_R \tag{2.63}$$

Azymuthal component, unit vector **m***:*

$$\frac{\partial v_\phi}{\partial t} + v_R \frac{\partial v_\phi}{\partial R} + \frac{v_R\, v_\phi}{R} + \frac{v_\phi}{R} \frac{\partial v_\phi}{\partial \phi} + v_z \frac{\partial v_\phi}{\partial z} = -\frac{1}{\rho\, R} \frac{\partial P}{\partial \phi} + \frac{1}{\rho} F_\phi \quad (2.64)$$

Vertical component, unit vector **k***:*

$$\frac{\partial v_z}{\partial t} + v_R \frac{\partial v_z}{\partial R} + \frac{v_\phi}{R} \frac{\partial v_z}{\partial \phi} + v_z \frac{\partial v_z}{\partial z} = -\frac{1}{\rho} \frac{\partial P}{\partial z} + \frac{1}{\rho} F_z \,. \quad (2.65)$$

2.4.1   Example: Azymuthal Symmetry

Assuming that v_ϕ and its derivatives are zero, there are only two equations:

$$\frac{\partial v_R}{\partial t} + v_R \frac{\partial v_R}{\partial R} + v_z \frac{\partial v_R}{\partial z} = -\frac{1}{\rho} \frac{\partial P}{\partial R} + \frac{1}{\rho} F_R \quad (2.66)$$

$$\frac{\partial v_z}{\partial t} + v_R \frac{\partial v_z}{\partial R} + v_z \frac{\partial v_z}{\partial z} = -\frac{1}{\rho} \frac{\partial P}{\partial z} + \frac{1}{\rho} F_z \,. \quad (2.67)$$

2.4.2   Example: The Unidimensional Case

We can imagine that the galactic disk is an infinitely large cylinder, so that its physical properties vary only with the height relative to the reference plane, and there is only one equation that can be written as

$$\frac{\partial v}{\partial t} + v \frac{\partial v}{\partial z} = -\frac{1}{\rho} \frac{\partial P}{\partial z} + \frac{1}{\rho} F \,, \quad (2.68)$$

where we dropped the subscript z.

2.5   Euler Equation in a Stellar Envelope

Let us consider a stellar envelope with spherical symmetry and in steady state, taking into account the stellar gravitational force. From (2.43), we can write for the gravitational force per unit volume

$$\begin{cases} \mathbf{F} = F_r\, \mathbf{n} \\ F_r = F = -\dfrac{G\, M_*\, \rho}{r^2} \,, \end{cases} \quad (2.69)$$

where M_* is the mass of the central object. Substituting in (2.53):

$$
\begin{cases}
v \dfrac{dv}{dr} = -\dfrac{1}{\rho}\dfrac{dP}{dr} - \dfrac{G\,M_*}{r^2} \\[2mm]
\rho\, v \dfrac{dv}{dr} = -\dfrac{dP}{dr} - \dfrac{G\,M_*\,\rho}{r^2} \; .
\end{cases}
\tag{2.70}
$$

Calling g_* the stellar gravitational acceleration,

$$
g_* = \frac{G\,M_*}{r^2} \; ,
\tag{2.71}
$$

we obtain

$$
\begin{cases}
v \dfrac{dv}{dr} = -\dfrac{1}{\rho}\dfrac{dP}{dr} - g_* \\[2mm]
\rho\, v \dfrac{dv}{dr} = -\dfrac{dP}{dr} - \rho\, g_* \; .
\end{cases}
\tag{2.72}
$$

We will be interested in problems where, apart from the gravitational attraction g_*, there is another external force per unit mass in the same direction but in opposition to the gravitational force. Such force may be, for instance, due to momentum transfer from the stellar radiation field to the gas in an expanding envelope. Calling g_r the acceleration (force per unit mass) due to this process, we have in the spherically symmetric and stationary case:

$$
\begin{aligned}
v \frac{dv}{dr} &= -\frac{1}{\rho}\frac{dP}{dr} - g_* + g_r \\[2mm]
&= -\frac{1}{\rho}\frac{dP}{dr} - (g_* - g_r) \\[2mm]
&= -\frac{1}{\rho}\frac{dP}{dr} - g_{ef} \; ,
\end{aligned}
\tag{2.73}
$$

where we have introduced the effective gravity

$$
g_{ef} = g_* - g_r \; .
\tag{2.74}
$$

Defining the Γ *parameter* by the ratio

$$
\Gamma_r = \frac{g_r}{g_*} \; ,
\tag{2.75}
$$

we may write

$$
g_{ef} = g_* \left(1 - \frac{g_r}{g_*} \right) = g_* \, (1 - \Gamma_r)
\tag{2.76}
$$

and the equation of motion (2.73) is written as

$$v \frac{dv}{dr} = -\frac{1}{\rho} \frac{dP}{dr} - g_* (1 - \Gamma_r) .$$

(2.77)

2.5.1   Example: Envelope in Hydrostatic Equilibrium

The simplest case of Eq. (2.77) is the hydrostatic equilibrium under the action of gravity. In this case, $\Gamma_r = 0$ e $v = 0$ in (2.77), so that

$$\frac{1}{\rho} \frac{dP}{dr} = -g_* = -\frac{G M_*}{r^2}$$

or

$$\frac{dP}{dr} = -\rho g_* = -\frac{G M_* \rho}{r^2} ,$$

(2.78)

which is the well-known hydrostatic equilibrium equation generally used in stellar structure theory. According to (2.78), the stellar fluid is in equilibrium if the pressure forces, characterized by the term dP/dr, exactly match the gravitational force. Since the product $\rho g_* > 0$, we must have $dP/dr < 0$, that is, in order to attain equilibrium, the pressure must decrease away from the center of the star.

2.5.2   Example: Hydrostatic Solar Corona

The solar atmospheric layers include a photosphere, a transition region and a hot corona, where the temperature may reach values $T \simeq 1.5 \times 10^6$ K with a density of protons and electrons $n \simeq 4 \times 10^8$ cm^{-3}. This hot gas expands through the interplanetary space, which essentially constitutes the *solar wind*, and the temperature decreases at large distances from the Sun. We can assume that the expansion is initiated at a reference distance given by $r_0 \simeq R_\odot = 6.96 \times 10^{10}$ cm. Let us make the simplifying assumption that the corona is in hydrostatic equilibrium and that the plasma is isothermal, so that the Earth, located at 1 AU $= 1.5 \times 10^{13}$ cm from the Sun, is inside a high temperature coronal material. We can use the hydrostatic equilibrium hypothesis in order to estimate the coronal gas density at the Earth's position. From (2.78), the hydrostatic equilibrium equation can be written

$$\frac{dP}{dr} = -\frac{G M_\odot n m_H}{r^2} .$$

(2.79)

Assuming that the plasma contains the same number of protons and electrons, we can write the equation of state in the form

$$P = 2 n k T , \tag{2.80}$$

where $k = 1.38 \times 10^{-16}$ erg/K is Boltzmann constant (see Chap. 3). From (2.79) and (2.80), we have

$$\frac{dP}{dr} = 2 k T \frac{dn}{dr} = -\frac{G M_\odot n m_H}{r^2} , \tag{2.81}$$

so that

$$\frac{dn}{n} = -\frac{G M_\odot m_H}{2 k T} r^{-2} dr = -\frac{1}{h} \left(\frac{r_0}{r} \right)^2 dr , \tag{2.82}$$

where we have introduced the *scale height h* given by

$$\frac{1}{h} = \frac{G M_\odot m_H}{2 k T r_0^2} . \tag{2.83}$$

Integrating (2.82) between r_0 and r, where the particle density has values n_0 and n, respectively, we obtain

$$n(r) = n_0 \exp\left[-\frac{r_0^2}{h} \left(\frac{1}{r_0} - \frac{1}{r} \right) \right] . \tag{2.84}$$

Taking $n_0 \simeq 4 \times 10^8$ cm^{-3} and $r_0 \simeq 6.96 \times 10^{10}$ cm, we get $h \simeq 9.0 \times 10^9$ cm $= 9.0 \times 10^4$ km. Therefore, at $r = 1$ AU the coronal gas density is $n(1 \text{ AU}) \simeq 2 \times 10^5$ cm^{-3}. In fact, the constant temperature hypothesis in the whole corona is unrealistic, and a better approximation gives lower values for the density at $r = 1$ AU.

2.5.3   Example: Free-Falling Stellar Envelope

In Example 2.5.1, we have seen that a negative pressure gradient is necessary $(dP/dr < 0)$ in order to attain hydrostatic equilibrium. Let us consider the case where the gradient goes abruptly to zero, leaving the gas in the envelope under the action of the gravity alone. From (2.77) we have

$$v \frac{dv}{dr} \simeq -g_* , \tag{2.85}$$

Fig. 2.4 Stellar envelope in
free fall from an initial
position to the stellar radius R

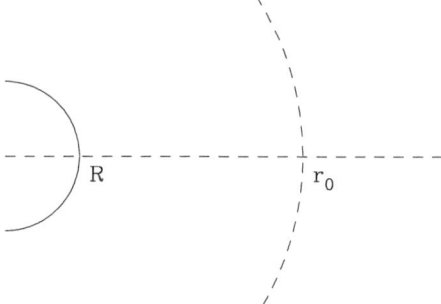

so that the gas in the envelope falls onto the star in free fall. Considering an initial
point at distance r_0 from the star, where $v_0 = 0$, let us obtain the velocity of the
fluid element as it approaches the stellar surface, where $r = R$ (Fig. 2.4).

Integrating (2.85) and assuming $g_* \simeq G\, M_*/R^2 \simeq$ constant, where M_* is the
stellar mass, we find

$$\int_0^{v(R)} v\, dv = -\int_{r_0}^{R} g_*\, dr \simeq g_*(r_0 - R)$$

(see Exercise 2.1). The result is well-known,

$$v(R)^2 \simeq 2\, g_*(r_0 - R)\,, \tag{2.86}$$

and the velocity is directed towards the stellar center. The time needed for the gas
initially at r_0 to fall on the stellar surface can be obtained from Eq. (2.77), recalling
that, in this case, $v\, dv/dr = v(dv/dt)/(dr/dt) = dv/dt$, and

$$\frac{dv}{dt} \simeq -g_*\,, \tag{2.87}$$

from which we have

$$t \simeq \frac{v(R)}{g_*}\,. \tag{2.88}$$

Considering the case of a red giant star with $M = 1\, M_\odot$, $R = 100\, R_\odot$ and $r_0 \simeq
2\, R$, we have $v(R) \simeq 62$ km/s and $t \simeq 26$ days, that is, the collapse of the envelope
in these conditions would be relatively fast.

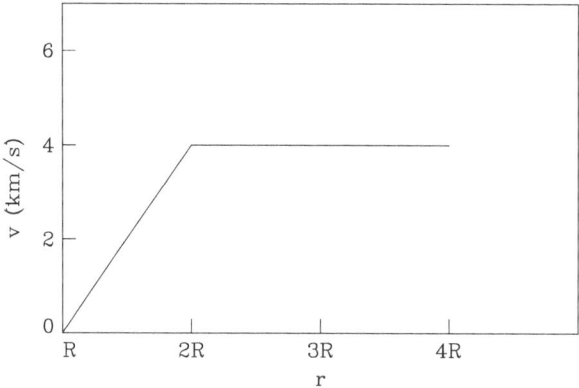

Fig. 2.5 Schematic velocity profile for an accelerated envelope

2.5.4   Example: Accelerated Envelope

Let us now consider the opposite situation of the previous example, in which the gravitational term is negligible compared to the pressure forces in a region of the envelope characterized by the width Δr. In this case, from (2.77),

$$v \frac{dv}{dr} \simeq -\frac{1}{\rho}\frac{dP}{dr} \ . \tag{2.89}$$

Considering that the average acceleration is simply given by $a \simeq |dP/dr|/\rho$, the gas is accelerated until it reaches the final velocity

$$v_f^2 \simeq 2 \left| \frac{dP}{dr} \right| \frac{\Delta r}{\rho} \ . \tag{2.90}$$

We can roughly estimate the final velocity in the circumstellar envelope of a red giant star using the equation of state (see Eq. 2.80), so that

$$v_f^2 \simeq 2 \frac{P}{\rho} \simeq 2 \frac{kT}{m_H} \ , \tag{2.91}$$

where $T \simeq 10^3$ K is the average temperature in the envelope, and $v_f \simeq 4$ km/s. This relation is simply a rough estimate, useful in order to obtain orders of magnitude. In contrast with Example 2.5.2, the gas is neutral, and the protons and electrons are locked in the hydrogen atoms or even in molecules. The behaviour of $v(r)$ is shown schematically by Fig. 2.5 for $\Delta r \simeq R$.

The observed winds in cool stars have profiles qualitatively similar to the figure. A variation of the velocity $v(r)$ as shown in Fig. 2.5 is known as a *velocity law*. The study of stellar winds frequently leads to relations of the type

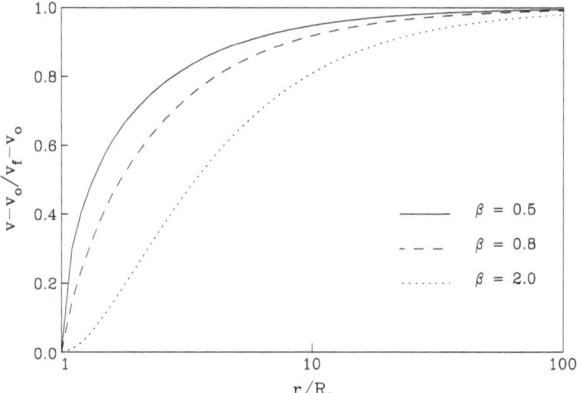

Fig. 2.6 Velocity profile for a beta law

$$v(r) = v_0 + (v_f - v_0) \left(1 - \frac{R}{r}\right)^{\beta}, \tag{2.92}$$

which is sometimes called a β *law*, where v_0 is the initial velocity of the wind, still in the photosphere, v_f is the final, or terminal velocity $(v_f \gg v_0)$, R is a reference level, usually of the order of the stellar radius R_*, and β is a parameter that characterizes the acceleration of the envelope. We see that $v \to v_0$ for $r \to R$ and that $v \to v_f$ for $r \to \infty$. In the case of hot stars, $\beta \simeq 0.8$, while for cool stars we have typically $\beta \simeq 2$. Some examples of velocity laws for different values of the β parameter are shown in Fig. 2.6, which is considerably more realistic than the schematic variation shown in Fig. 2.5.

2.5.5   Example: The Escape Velocity

We have seen in Examples 1.7.2 and 1.7.3 that the winds in hot stars have terminal velocities of the order of $v_f \sim 2{,}000\,\text{km/s}$, while the winds observed in red giant stars are considerably slower, with $v_f \sim 10\,\text{km/s}$. We can compare these values with the *escape velocity* v_e from the atmospheres of these stars. Considering a particle of mass m and velocity v at a distance r from the center of a star with mass M_* and radius R_*, the escape velocity is defined by the condition $E_c = |E_p|$, where $E_c = (1/2)mv^2$ is the kinetic energy of the particle and $|E_p| = GM_*m/r$ is its potential energy at r, where we have $r \geq R_*$. The escape velocity at this position is then

$$v_e \simeq \sqrt{\frac{2\,G\,M_*}{r}}. \tag{2.93}$$

For a B-type supergiant, adopting $M_* \simeq 25\,M_\odot \simeq 5 \times 10^{34}$ g and $r \simeq R_* \simeq 30\,R_\odot \simeq 2 \times 10^{12}$ cm, we have $v_e \simeq 600$ km/s, that is, the terminal velocities obtained from the observations are higher than the escape velocity, $v_f \gg v_e$.

In the case of the red giants, the situation is somewhat more complex. Taking typical values for these stars, $M_* \simeq 1\,M_\odot \simeq 2 \times 10^{33}$ g and $R_* \simeq 100\,R_\odot \simeq 7 \times 10^{12}$ cm, we get $v_e \simeq 60$ km/s with $r \simeq R_*$, that is, in this case $v_f < v_e$. This question was solved from the observations of the α Her system by Armin Deutsch. In this system, a red giant star and a dwarf star rotate around their center of mass. The circumstellar spectral lines that characterize the red giant outer atmosphere were also observed in the companion star, that is, both stars were inside a giant envelope, originated by the giant star. In this way, the position r where we should apply the condition $E_c = |E_p|$ occurs at a much larger distance than the stellar radius, $r \gg R_*$, so that the escape velocity from the envelope is in fact lower than the value given above. For example, including in (2.93) the masses of both stars, assumed identical, the escape velocity $v_e = v_f = 10$ km/s is obtained at position $r \simeq 2GM_*/v_e^2 \simeq 5 \times 10^{14}$ cm, or about 70 stellar radii. The observed dimensions of the circumstellar envelopes are effectively of this order of magnitude, or even larger.

2.5.6 Example: The Γ Parameter in Hot Stars

We can obtain an estimate of the Γ parameter defined by (2.75) in the envelopes of hot stars, where the large electron density makes electron scattering an important source of opacity. We know that the cross section for electron scattering in the non relativistic case is simply the Thomson cross section, $\sigma_T = 6.65 \times 10^{-25}$ cm^2. In a layer of density ρ and electron density n_e, the electron scattering coefficient per mass is

$$\kappa_e \simeq \frac{n_e\,\sigma_T}{\rho}\,, \tag{2.94}$$

typically of the order of $\kappa_e \simeq 0.30$ cm^2/g for a gas with normal chemical composition. Each absorbed (or scattered) photon with energy $h\nu$ has momentum $h\nu/c$, so that the ratio between the acceleration due to the scattering process and gravity is simply

$$\Gamma_e = \frac{g_e}{g_*} = \frac{L_*\,\kappa_e}{4\,\pi\,r^2\,c}\,\frac{r^2}{G\,M_*} = \frac{L_*\,\kappa_e}{4\,\pi\,c\,G\,M_*}\,, \tag{2.95}$$

where L_* is the stellar luminosity (erg/s), $L_*/4\pi r^2$ is the stellar flux at distance r (erg cm^{-2} s^{-1}), so that the term $L_*\kappa_e/4\pi r^2 c$ is essentially the acceleration (force/mass) due to the electron scattering process. For solar-type stars, Γ_e is very small, $\Gamma_e \simeq 2 \times 10^{-5}$, but using appropriate values for hot stars of O, B spectral types, $L_* \sim 10^5$ to $10^6\,L_\odot$ and $M_* \simeq 10$ to $50\,M_\odot$, respectively, we obtain

$0.1 \leq \Gamma_e \leq 1$, that is, in these stars the acceleration due to electron scattering may be similar to the gravitational acceleration. Parameter Γ_e is called the Eddington parameter, and $\Gamma_e = 1$ defines the Eddington limit, in which case the star is gravitationally unbound. From (2.95) we have at the Eddington limit

$$\frac{L_* \kappa_e}{4 \pi c G M_*} = 1 . \tag{2.96}$$

Recalling that for a spherical star $L_* = 4 \pi R_*^2 \sigma T_{eff}^4$ (see Eq. 1.31) and $g(R_*) = G M_*/R_*^2$, we have

$$g(R_*) = \frac{\sigma \kappa_e}{c} T_{eff}^4 \tag{2.97}$$

that is, plotting $\log g$ against $\log T_{eff}$ we get a straight line above which the star is gravitationally unbound,

$$\log g \simeq -15.25 + 4 \log T_{eff} . \tag{2.98}$$

For example, for $T_{eff} = 30,000\,\mathrm{K}$ we have $\log g = 2.66$, or $g \simeq 460\,\mathrm{g/cm}^2$.

In terms of the Γ_e parameter, we can define an effective escape velocity given by

$$v_e = \left[\frac{2 (1 - \Gamma_e) G M_*}{r} \right]^{1/2} . \tag{2.99}$$

Using data from Example 2.5.5, $M_* \simeq 25\,M_\odot$ and $r \simeq 30\,R_\odot$, taking $L_* \simeq 4 \times 10^5\,L_\odot$ and $\kappa_e \simeq 0.30\,\mathrm{cm}^2/\mathrm{g}$, we get $\Gamma_e \simeq 0.37$; from (2.99), we find an effective escape velocity corresponding to 80 % of the value given by (2.93).

2.5.7   Example: The Γ Parameter in Cool Stars

We can use the same procedure of the previous example to estimate the Γ parameter in the case of cool giant stars. Considering that the contribution g_r to the effective gravity is due to the action of the stellar radiation on the dust grains embedded in the envelope (see Eqs. 2.74 and 2.75), Eq. (2.95) is still valid, replacing the scattering coefficient κ_e by the corresponding quantity for solid grains, κ_d, given by

$$\kappa_d \simeq \frac{\pi a^2 Q n_d}{\rho} , \tag{2.100}$$

where n_d the grain number density (cm^{-3}), assuming spherical grains with radius a, and Q is an efficient factor for radiation pressure. Since $n_d = \rho_d/m_d$, where ρ_d is the grain density $(\mathrm{g/cm}^3)$ and $m_d \simeq (4/3)\pi a^3 s_d$ is the mass of a grain with internal density s_d, we can use average values adequate to silicate grains and estimate κ_d or the corresponding Γ parameter by

$$\Gamma_d = \frac{L_* \kappa_d}{4 \pi c G M_*} \,. \tag{2.101}$$

These values are: $a \simeq 1{,}000\,\text{Å} = 10^{-5}$ cm, $s_d \simeq 3\,\text{g/cm}^3$ and $\rho/\rho_d \simeq 200$ (gas-grain ratio). We get $m_d \simeq 1.3 \times 10^{-14}$ g, $n_d/\rho \simeq 1/200\,m_d \simeq 3.9 \times 10^{11}\,\text{g}^{-1}$ and $\kappa_d \simeq 120\,Q\,\text{cm}^2/\text{g}$. Coefficient κ_d depends critically on the photon frequency, reflecting the dependency of the efficiency factor Q, which can reach values from about 0.001 up to unity, approximately. Taking values in the range $0.01 < Q < 1$ and considering a star with $L_* \simeq 10^3\,L_\odot$ and $M_* \simeq 1 M_\odot$, from (2.101) we get $0.1 < \Gamma_d < 9$, that is, even in this case the added gravity term can be of the same order or higher than the stellar gravity.

2.5.8 Example: Expanding Planetary Nebula

An application of Example 2.5.4, where the gravitational attraction is not important, occurs in planetary nebulae, in which the envelope expands at an approximately constant velocity in the regions far away from the central star. For instance, a proton in the ionized region is typically located at a distance $r \simeq 0.2$ pc from the star. Considering an average mass $M_* \simeq 1\,M_\odot$, the proton potential energy at this position is $|E_p| \simeq G\,M_*\,m_p/r \simeq 3.6 \times 10^{-16}$ erg. Taking typical expansion velocities for planetary nebulae, $v \simeq 20$ km/s, the proton kinetic energy is $E_c \simeq (1/2)\,m_p\,v^2 \simeq 3.3 \times 10^{-12}$ erg, so that $E_c \gg |E_p|$, and the envelope escapes from the star. In other words, the expansion velocity is higher than the gas escape velocity, which is given by (2.93). In this case, $v_e \simeq 0.2$ km/s, or $v \gg v_e$.

Exercises

2.1. Solve the case of Example 2.5.3 releasing the assumption that g_* is a constant. Show that in this case $v(R) \simeq 62/\sqrt{2} \simeq 44$ km/s. What is the time needed for collapse?

2.2. Considering that the Sun is a spherical star with mass $M_\odot = 1.99 \times 10^{33}$ g and radius $R_\odot = 6.96 \times 10^{10}$ cm, estimate the value of the pressure gradient in the Sun. Compare your result with the value obtained by the so-called "solar standard model" for the region in the solar interior where $r = R_\odot/2$, which is $dP/dr \simeq -1.3 \times 10^5\,\text{dyn/cm}^3$.

2.3. In a more realistic model for the solar corona, the temperature variation is given by $T(r) = T_0\,(r_0/r)^{2/7}$, where $r_0 \simeq R_\odot = 6.96 \times 10^{10}$ cm and $T_0 \simeq 1.5 \times 10^6$ K are the values of the position r and temperature T at a certain reference level near the surface of the Sun. (a) What is the temperature of the coronal gas at the Earth's orbit? (b) Assume that the solar corona is in hydrostatic equilibrium. What is the variation of the coronal gas density with distance? What is the value of the density at $r = 1\,\text{AU}$?

2.4. Consider a β velocity law for a stellar wind in which the reference level R is related to the stellar radius R_* by

$$R = R_* \left[1 - \left(\frac{v_0}{v_f} \right)^{\frac{1}{\beta}} \right]$$

(a) What is the value of R, in terms of the stellar radius, for a hot star of O spectral type, with effective temperature $T_{eff} = 40{,}000\,\text{K}$, and a wind with terminal velocity $2{,}500\,\text{km/s}$ and $\beta = 0.8$? Assume that the initial wind velocity is essentially given by the sound speed at the base of the circumstellar envelope, that is, $v_0^2 \simeq c_s^2 \simeq k\,T_{eff}/\mu\,m_H$, where $\mu \simeq 0.6$ is the mean molecular weight of the gas particles. (b) Plot the velocity $v(r)$ as a function of r/R_*. At what distance from the star, in terms of R_*, the wind reaches 60 % of the terminal velocity?

2.5. Hot stars with O, B spectral types and effective temperatures higher than $21{,}000\,\text{K}$ present a ratio between the wind terminal velocity v_f and the effective escape velocity v_e given by $v_f/v_e \simeq 2.6$. The star ζ Pup has O4f spectral type, effective temperature $T_{eff} \simeq 42{,}000\,\text{K}$, mass $M_* \simeq 60\,M_\odot$, radius $R_* \simeq 20\,R_\odot$ and a wind with terminal velocity $v_f \simeq 2{,}200\,\text{km/s}$. (a) What is the average value of the Γ parameter for this star, defined as the ratio between the acceleration due to electron scattering and the stellar gravitational acceleration? (b) Considering that the total stellar luminosity is $L_* \simeq 8 \times 10^5\,L_\odot$, what is the electron scattering coefficient (cm^2/g) in the stellar envelope?

Bibliography

Apart from the references given in Chap. 1, in particular the books by Landau and Lifchitz (1971), Batchelor (1967), and Clarke and Carswell (2007), other texts can be used, as follows.

Arfken, G.B., Weber, H.J., Harris, F.E.: Mathematical Methods for Physicists. Academic, New York (2012) (New Edition of a Classic Book on Mathematical Methods of Physics)

Choudhuri, A. R.: The Physics of Fluids and Plasmas. Cambridge University Press, Cambridge (1998) Referred to in Chapter 1. Includes a good discussion of the Euler equation, especially in MHD

Castor, J.I., Abbott, D.C., Klein, R.I.: Astrophys. J. **195**, 157 (1975) (Classic paper on Stellar Winds in Hot Stars, with a Discussion of the β Velocity Law. See Also Astrophys. J. Suppl. Ser. **139**, 197 (1975))

Munson, B.R., Rothmayer, A.P., Okiishi, T.H., Huebsch, W.W.: Fundamentals of Fluid Mechanics. Wiley, New York (2012) (Introductory Text on Fluid Mechanics, Directed to Engineering Students, with Emphasis on the Fundamental Aspects of Fluid Flows, Including Eulerian and Lagrangian Descriptions)

Symon, K.R.: Mechanics. Addison-Wesley, Reading (1971) (Basic mechanics book, with a discussion on vectors and coordinates)

Chapter 3
The Equation of State

Abstract This chapter introduces the equation of state, defines the mean molecular weight, discusses the validity of the perfect gas hypothesis, and presents several examples regarding stellar photospheres.

3.1 Introduction

The state of a moving ideal gas was characterized until now by the velocity $\mathbf{v}(\mathbf{r}, t)$, density $\rho(\mathbf{r}, t)$ and pressure $P(\mathbf{r}, t)$, which are given as functions of the position $\mathbf{r}$ and time t. These quantities obey the general conservation laws, namely conservation of *mass*, *momentum* and *energy*. Since the equation of motion is a vector equation, the state of the gas involves five scalar equations, in terms of the variables P, ρ, v_x, v_y and v_z, for example. Naturally, the remaining quantities that appear in the hydrodynamics equations must be known, or related to thermodynamic properties of the gas. For example, the specific internal energy may be determined either in terms of the pressure and density, $e(P, \rho)$, or in terms of the absolute temperature, $e(T)$. The latter is related to the remaining thermodynamic quantities by the *equation of state*, which can be written in the form

$$f(P, \rho, T) = 0 \,. \tag{3.1}$$

In Examples 2.5.2 and 2.5.4 we have used Eqs. (2.80) and (2.91), which are of the form $P \propto \rho\, T$, that is, they are forms of the equation of state.

W.J. Maciel, *Hydrodynamics and Stellar Winds: An Introduction*, Undergraduate
Lecture Notes in Physics, DOI 10.1007/978-3-319-04328-9_3,
© Springer International Publishing Switzerland 2014

3.2   The Equation of State of a Perfect Gas

For a non-degenerate, ideal gas, the equation of state is

$$P V = v \mathscr{R} T , \qquad (3.2)$$

where V is the volume occupied by the gas, $\mathscr{R} = 8.31 \times 10^7 \, \text{erg mole}^{-1} \, \text{K}^{-1}$ is the gas constant and v is the number of moles,

$$v = \frac{N}{N_a} , \qquad (3.3)$$

where N is the number of gas particles and $N_a = 6.02 \times 10^{23}$ molecules/mole is Avogadro's number. Considering (3.3), Eq. (3.2) is written as

$$P V = N \, \frac{\mathscr{R}}{N_a} \, T . \qquad (3.4)$$

Since

$$\mathscr{R} = N_a \, k , \qquad (3.5)$$

where $k = 1.38 \times 10^{-16}$ erg/K is Boltzmann's constant, we may write

$$P V = N k T . \qquad (3.6)$$

Using the particle number density,

$$n = \frac{N}{V} , \qquad (3.7)$$

we obtain

$$P = n k T . \qquad (3.8)$$

Calling m the particle mass, we have for a homogeneous gas

$$\rho = n \, m \qquad (3.9)$$

and therefore

$$P = \frac{k \, \rho \, T}{m} . \qquad (3.10)$$

3.3   The Mean Molecular Weight

Defining the mean molecular mass, or "molecular weight" for a homogeneous gas,

$$\mu = \frac{m}{m_H} \, , \tag{3.11}$$

where $m_H = 1.67 \times 10^{-24}$ g is the H atom mass, we have

$$P = \frac{k \rho T}{\mu \, m_H} \, , \tag{3.12}$$

which is the usual form of the equation of state. If the gas is not homogeneous, that is, in the case of a mixture of different particles with mass m_i at a temperature T, Eq. (3.12) is still valid, as long as

$$\rho_i = n_i \, m_i \tag{3.13}$$

$$\rho = \sum \rho_i = \sum n_i \, m_i \, . \tag{3.14}$$

The *mean molecular weight* is defined as

$$\mu = \frac{1}{m_H} \frac{\sum n_i \, m_i}{\sum n_i} \, . \tag{3.15}$$

Strictly speaking, the unit of mass used in (3.11) or (3.15) is the atomic mass unit, adopted as $1/12$ of the mass of the ^{12}C nucleus. In practical terms, however, we may consider this mass as essentially equal that of the hydrogen atom m_H.

In the case of a mixture of particles, comparing equations (3.11) and (3.15) we see that the former is still valid, as long as the mass m is understood as the mean mass per particle, or $m = \sum n_i \, m_i / \sum n_i$. Therefore, in this case μ is the mean molecular weight per free particle, that is, increasing the number of free particles (for instance by ionization), but keeping the total gas mass constant, the mean molecular weight μ decreases.

3.3.1   Example: Neutral H Gas

In this case, the gas contains only H I (or H°), and the mean molecular weight is

$$\mu = \frac{1}{m_H} \frac{n_{HI} \, m_H}{n_{HI}} = 1 \, . \tag{3.16}$$

3.3.2 Example: Neutral H Gas with 10 % of Neutral He

The mean molecular weight can be written as

$$\mu = \frac{1}{m_H} \frac{n_{HI}\, m_H + n_{HeI}\, m_{He}}{n_{HI} + n_{HeI}} = \frac{1}{m_H} \frac{n_{HI}\, m_H + 4\, n_{HeI}\, m_H}{n_{HI} + n_{HeI}}$$

$$\mu = \frac{1 + 4(n_{HeI}/n_{HI})}{1 + (n_{HeI}/n_{HI})} = \frac{1 + 0.4}{1 + 0.1} = 1.27 \tag{3.17}$$

where we have used $n_{HeI}/n_{HI} = 0.10$.

3.3.3 Example: Completely Ionized H Gas

In this case, the gas contains H II (or H$^+$), and the mean molecular weight is

$$\mu = \frac{1}{m_H} \frac{n_{HII}\, m_H + n_e\, m_e}{n_{HII} + n_e} \simeq \frac{1}{m_H} \frac{n_{HII}\, m_H}{2\, n_{HII}} \simeq 0.5, \tag{3.18}$$

where we have used the fact that the electron density is $n_e = n_{HII}$ and $n_e\, m_e \ll n_{HII}\, m_H$, since the electron mass $m_e = 9.11 \times 10^{-28}$ g is much smaller than m_H.

3.3.4 Example: Ionized H Gas with 10 % Neutral He

In this case, we have $n_e = n_{HII}$ and $n_{HeI}/n_{HII} = 0.10$, so that the mean molecular weight is

$$\mu = \frac{1}{m_H} \frac{n_{HII}\, m_H + n_e\, m_e + n_{HeI}\, m_{He}}{n_{HII} + n_e + n_{HeI}} \simeq \frac{1}{m_H} \left(\frac{n_{HII} + 4\, n_{HeI}}{2\, n_{HII} + n_{HeI}} \right) m_H$$

$$\mu \simeq \frac{1 + 4(n_{HeI}/n_{HII})}{2 + (n_{HeI}/n_{HII})} \simeq \frac{1 + 4(0.1)}{2 + 0.1} \simeq 0.67. \tag{3.19}$$

3.3.5 Example: Molecular Hydrogen Gas

The mean molecular weight is now

$$\mu = \frac{1}{m_H} \frac{n_{H_2}\, m_{H_2}}{n_{H_2}} = 2. \tag{3.20}$$

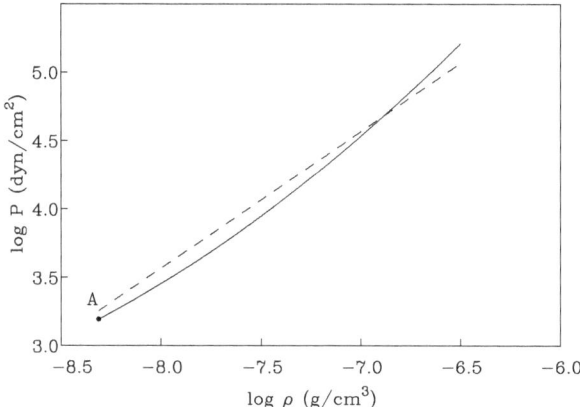

Fig. 3.1 Variation of the pressure in the solar photosphere based on a numerical model (*solid line*) and the approximation given by (3.12) (*dashed line*)

3.3.6 Example: The Solar Photosphere

We can obtain a better idea of the application of the ideal gas equation of state (3.12) considering the variation of the main physical quantities in the solar photosphere. A numerical model for the solar photosphere, assuming an effective temperature $T_{eff} = 5{,}800\,\mathrm{K}$ and a gas mixture containing H and 10 % of He, leads to a variation between the gas pressure and density as shown by the solid line in Fig. 3.1. In this case, the lower values of the pressure and density correspond approximately to the coolest photospheric regions, represented by point A in the figure. The figure does not show the regions where the inversion of the temperature occurs, which are characteristic of the solar chromosphere, transition region and corona. These regions are located further left of point A.

The photospheric region shown in the figure has a total spatial variation of about 600 km (recall that the solar radius is $R_{\odot} = 6.96 \times 10^{10}\,\mathrm{cm}$), in which the pressure changes by a factor of 150, while the density varies by a factor of 70. On the other hand, the temperature changes by a factor lower than 2, from the outer layers to the region characterized by $T \simeq T_{eff}$. Let us then neglect this variation in a first approximation, and consider $T \simeq$ constant $\simeq T_{eff} = 5{,}800\,\mathrm{K}$. In this case, we can use (3.12) to estimate the photospheric pressure, adopting $\mu \simeq 1.3$. The result is shown by the dashed line in Fig. 3.1. We see that the error from the hypothesis of a constant temperature is small, of the order of 30 %, decreasing for the regions where the temperature is near the effective temperature, where the two lines of Fig. 3.1 are closer to each other.

3.4   Validity of the Perfect Gas Hypothesis

A perfect gas can be characterized by the fact that the average interaction energy of the particles E_i is much smaller than the thermal energy E_t, that is

$$E_i \ll E_t . \tag{3.21}$$

This condition is fulfilled when the interaction among the particles is small, or the gas is sufficiently rarefied. Let us consider some simple examples applied to astrophysical situations, particularly to expanding stellar envelopes, and show that in general the perfect gas hypothesis is a reasonable one, so that the ideal gas equation of state can be used as a first approximation.

The interaction potential energy E_i in (3.21) increases as the particles are closer to each other, that is, when the gas density increases. Moreover, different physical processes have different values of the interaction energy at a given interparticle distance. For example, in the case of electrical interactions among the gas particles, the validity of the ideal fluid hypothesis can be examined by the comparison of the coulomb interaction energy E_{coul} with the gas thermal energy E_t. The former can be written as:

$$E_{coul} \sim \frac{e^2}{r} , \tag{3.22}$$

where $e = 4.80 \times 10^{-10}\,\mathrm{g}^{1/2}\,\mathrm{cm}^{3/2}\,\mathrm{s}^{-1}$ is the electron charge and r is the average separation of the particles. On the other hand, the thermal energy is given by

$$E_t \sim \frac{3}{2} k\,T \tag{3.23}$$

where T is the gas kinetic temperature. To keep the validity of the ideal fluid hypothesis, we must have in general $E_{coul} \ll E_t$. If this condition is not satisfied, there is a classical deviation from the perfect gas, which can be taken into account by the use of a modified equation of state, such as the van der Waals equation. However, in the astrophysical applications considered here this is almost never necessary. Other non-classical deviations may be important in the denser regions of the stellar interior, such as the electron degeneracy. Possibly, the main interaction to be considered in some specific cases in the study of stellar winds is the presence of viscosity, which deeply affects the fluid equations, as we will see in Chap. 7.

3.4.1   Example: Neutral Gas Envelope

A non-ionized gas behaves as a perfect gas for densities lower than a certain critical value, of the order of $1\,\mathrm{g/cm}^3$ for "ordinary" matter. For densities higher than this value, the gas becomes compression resistant. This limitation is imposed by interatomic forces, which act at distances of the order of the atomic dimensions, $\sim 10^{-8}$ cm. In a neutral gas envelope, such as those around red giant stars, we can estimate the average separation of the gas particles considering $r \sim n^{-1/3}$, or $r \sim 10^{-3}$ cm with $n \sim 10^8$ cm^{-3}. Therefore, $E_{coul} \sim 10^{-16}$ erg. At a temperature $T \sim 10^3$ K, $E_t \sim 10^{-13}$ erg, or $E_{coul} \ll E_t$. Strictly speaking, the obtained value for E_{coul} is an upper limit, since for neutral atoms the coulomb interaction is less important, as the protons and electrons in each atom interact weakly with their neighbouring atoms, depending on the relative orientation of the electric charges. In a more correct expression for E_{coul}, the average separation r in (3.22) should be replaced by r^3/d^2, where $d \sim 10^{-8}$ cm, of the order of the radius of the first Bohr orbit. In this case, $E_{coul} \sim 10^{-26}$ erg, that is, $E_{coul} \ll E_t$.

3.4.2 Example: Ionized Gas Envelope

A completely ionized gas behaves as a perfect gas, even at relatively large densities. In ionized matter, the particles occupy dimensions of the order of the nuclear radius ($\sim 10^{-13}$ cm), that is, 10^5 times smaller than in non-ionized matter. Therefore, the volume occupied by the ionized gas is about 10^{15} times smaller than in the case of the non-ionized gas, which implies a larger compression possibility for the ionized gas. Considering a stellar envelope with $n \sim 10^4$ cm^{-3} and $T \sim 10^4$ K we have $r \sim 10^{-2}$ cm. From (3.22) and (3.23), $E_{coul} \sim 10^{-17}$ erg and $E_t \sim 10^{-12}$ erg, or $E_{coul} \ll E_t$. Comparing (3.22) and (3.23) we can conclude that the ideal gas hypothesis is valid for $T \gg 10^{-3} n^{1/3}$, where T is in K and n in cm^{-3}.

Exercises

3.1. Prove the perfect gas equation of state, Eq. (3.8). Hint: Consider the momentum transfer from the gas particles to an imaginary surface S. Assume thermodynamic equilibrium. so that the momentum distribution of the particles is given by the Maxwell-Boltzmann distribution function.

3.2. (a) Calculate the molecular weight of a gas made of ionized hydrogen with 10 % of once-ionized helium, that is, $n_{HeII}/n_{HII} = 0.10$. (b) What is the molecular weight if helium is twice ionized?

3.3. A planetary nebula composed of ionized H and He has an electron temperature $T_e = 10^4$ K and density $n_e = 10^4$ cm^{-3}. (a) What is the average pressure in the nebula? (b) Assuming that the nebula is expanding at a velocity of 20 km/s, what is the mass flux to the interstellar medium? (c) Considering that the nebula has an angular diameter of 1 arcmin and is located at 2 kpc from the Sun, determine its age, assumed equal to the expansion time.

3.4. Consider a red giant star with M5III spectral type, mass of $1\,M_\odot$, effective temperature of $3{,}000\,\mathrm{K}$ and surface gravity given by $\log g = 1.0$, where g is in $\mathrm{cm/s^2}$. A model atmosphere for the star leads to a total pressure given by $\log P = 3.5$, where the pressure is in ($\mathrm{dyn/cm^2}$), in a region where the temperature is essentially the same as the effective temperature. (a) What is the density in this region, if the molecular weight in the atmosphere is $\mu = 1.3$? (b) The star has a circumstellar envelope with $T = 10^3\,\mathrm{K}$ at about 5 stellar radii from the center, and the gas escapes with a terminal velocity of $20\,\mathrm{km/s}$ and a mass loss rate of $10^{-6}\,M_\odot/\mathrm{year}$. What is the average pressure in the envelope? Compare your result with a typical value for the pressure in the stellar atmosphere.

3.5. The temperature of the photospheric layers of a star with effective temperature $T_{eff} = 20{,}000\,\mathrm{K}$ increases about 50 % from the surface of the star to the region characterized by the effective temperature. (a) Considering that the photospheric density varies from $2.0\times10^{-11}\,\mathrm{g/cm^3}$ to $9.0\times10^{-10}\,\mathrm{g/cm^3}$ in the same region, what is the expected variation of the gas pressure? Adopt $\mu = 1.0$. (b) Considering that the pressure is essentially due to the electrons, what is the variation of the electron density in this region?

Bibliography

Cox, A.N.: Allen's Astrophysical Quantities. Springer, New York (2001) (Very useful compilation of tables and physical and astronomical data, constants, and conversion of units. Figure 3.1 was based on this reference)

Kippenhahn, R., Weigert, A., Weiss, A.: Stellar Structure and Evolution. Springer, Berlin (2012) (Second edition of a basic book on the structure and evolution of the stars, with a discussion of the basic equations of stellar structure, in particular the equation of state for perfect gases, classical and relativistic deviations)

Reif, F.: Fundamentals of Statistical and Thermal Physics. McGraw-Hill, New York (1965) [recent edition: Waveland Press (2008)] (Basic text on thermodynamics and statistical physics, with a good discussion of the thermodynamic equilibrium equations and the equation of state)

Chapter 4
The Energy Equation

Abstract This chapter introduces the energy equation both in the Eulerian and Lagrangian forms, and presents some simple applications in the adiabatic case.

4.1 Introduction

In the absence of magnetic fields, a non-viscous fluid can be characterized by five quantities, such as the three components of the velocity $\mathbf{v}$, the pressure P and the density ρ. The five equations that determine these quantities are the Euler equation (three components), the continuity equation and an equation that expresses the energy conservation in the fluid. Such description corresponds to the "continuous model", in which the hydrodynamic equations are written in terms of the macroscopic variables P, ρ and $\mathbf{v}$ as functions of the position $\mathbf{r}$. At a more fundamental level, the gas can also be described by a distribution function $f(\mathbf{r}, \mathbf{v}, t)$, and the variations of this function are given by *Boltzmann equation*, in the case of neutral fluids, or by *Vlasov equation* for plasmas.

The simplest case of the energy equation assumes an isothermal flow, that is, the temperature is considered as constant, so that the energy equation can be written as

$$T = \text{constant} . \tag{4.1}$$

In this chapter, we will consider some examples of the energy equation which are applicable to astrophysical situations in general and to stellar winds in particular. Before we do that, let us review the simple case of an adiabatic flow. More experienced readers may omit the next section.

W.J. Maciel, *Hydrodynamics and Stellar Winds: An Introduction*, Undergraduate Lecture Notes in Physics, DOI 10.1007/978-3-319-04328-9_4,
© Springer International Publishing Switzerland 2014

4.2   Adiabatic Motion

In an *isentropic* process, the entropy per unit mass s (with units $\mathrm{erg\,K^{-1}\,g^{-1}}$) is constant. This situation may occur in the motion of a perfect gas if there is no energy exchange among the different parts of the fluid, or between the fluid and the surrounding gas. In this case, the motion is *adiabatic*, and the entropy in every part of the fluid remains constant during the flow. We can write then

$$s = \text{constant} . \tag{4.2}$$

Let us obtain the equations for an adiabatic flow from basic principles. From the first law of thermodynamics,

$$dQ = dE + P\,dV , \tag{4.3}$$

where dQ is absorbed heat and dE is the variation in the internal energy of the fluid. In a moving fluid, such as in the case of an expanding stellar envelope,

$$dQ = 0 . \tag{4.4}$$

As the gas expands, it does work at the expense of its internal energy. Consequently, its absolute temperature T changes. From (4.3) and (4.4) we have

$$dE + P\,dV = 0 . \tag{4.5}$$

In an ideal gas, the internal energy depends only on the temperature,

$$E = E(T) . \tag{4.6}$$

Therefore,

$$dE = \left(\frac{\partial E}{\partial T} \right)_V dT . \tag{4.7}$$

Let us introduce the molar specific heat at constant volume:

$$c_V = \frac{1}{\nu} \left(\frac{dQ}{dT} \right)_V . \tag{4.8}$$

With $dV = 0$, we have the equations:

$$dQ = dE , \tag{4.9}$$

$$c_V = \frac{1}{\nu} \frac{dE}{dT} , \tag{4.10}$$

$$dE = \nu\, c_V \, dT . \tag{4.11}$$

From (4.11) and (4.3) with $dQ = 0$,

$$0 = \nu \, c_V \, dT + P \, dV \, . \tag{4.12}$$

The equation of state is

$$P V = \nu \, \mathscr{R} \, T \, . \tag{4.13}$$

From (4.13),

$$P \, dV + V \, dP = \nu \, \mathscr{R} \, dT \, ,$$

$$\nu \, dT = \frac{P}{\mathscr{R}} \, dV + \frac{V}{\mathscr{R}} \, dP \, .$$

Substituting in (4.12),

$$0 = c_V \, \frac{P}{\mathscr{R}} \, dV + c_V \, \frac{V}{\mathscr{R}} \, dP + P \, dV = (c_V + \mathscr{R}) \, P \, dV + c_V \, V \, dP = \frac{c_V + \mathscr{R}}{c_V} \, \frac{dV}{V} + \frac{dP}{P} \, . \tag{4.14}$$

Introducing the specific heat ratio

$$\gamma = \frac{c_P}{c_V} \, , \tag{4.15}$$

where c_P is the specific heat at constant pressure,

$$c_P = \frac{1}{\nu} \left(\frac{dQ}{dT} \right)_P \, . \tag{4.16}$$

From (4.3) and (4.11),

$$dQ = \nu \, c_V \, dT + P \, dV \, . \tag{4.17}$$

From (4.13), with constant P,

$$P \, dV = \nu \, \mathscr{R} \, dT \, .$$

Therefore,

$$dQ = \nu \, c_V \, dT + \nu \, \mathscr{R} \, dT \, . \tag{4.18}$$

But from (4.16) and (4.18)

$$c_P = c_V + \mathscr{R} \, . \tag{4.19}$$

From (4.15) and (4.19),

$$\gamma = \frac{c_V + \mathcal{R}}{c_V} = 1 + \frac{\mathcal{R}}{c_V} . \tag{4.20}$$

Using (4.20) in (4.14),

$$\begin{cases} \gamma \dfrac{dV}{V} + \dfrac{dP}{P} = 0 \\[2mm] \dfrac{dP}{P} = -\gamma \dfrac{dV}{V} \\[2mm] \dfrac{d\ln P}{d\ln V} = -\gamma . \end{cases} \tag{4.21}$$

In terms of the gas density ρ,

$$\frac{d\rho}{\rho} = -\frac{dV}{V} . \tag{4.22}$$

Using (4.22) and (4.21),

$$\begin{cases} \dfrac{dP}{P} = \gamma \dfrac{d\rho}{\rho} \\[2mm] \dfrac{d\ln P}{d\ln \rho} = \gamma . \end{cases} \tag{4.23}$$

Generally, c_P, c_V, and γ depend on the temperature T, although in many instances γ changes slowly with the temperature. If γ is considered as independent of T, Eqs. (4.21) and (4.23) can be integrated directly, and we have

$$\begin{cases} P = \text{constant} \times V^{-\gamma} \\ P = \text{constant} \times \rho^{\gamma} . \end{cases} \tag{4.24}$$

In terms of T, taking into account the equation of state, we have

$$\begin{cases} T = \text{constant} \times V^{1-\gamma} \\ T = \text{constant} \times \rho^{\gamma-1} . \end{cases} \tag{4.25}$$

It is interesting to compare the variation of the pressure P with volume V in the adiabatic and isothermal cases. For the latter we have

$$P V = \text{constant} , \tag{4.26}$$

whereas, from (4.24), $P V^{\gamma} = \text{constant}$. We have $c_P > c_V$, since $c_P = c_V + \mathcal{R}$, so that $\gamma = c_P/c_V > 1$, and in an adiabatic expansion the pressure changes more rapidly with the gas volume than in the isothermal case. In an isothermal flow, as the volume increases the pressure decreases, and the temperature is kept constant due to some external energy source. On the other hand, in the adiabatic flow, work is done at the expense of the gas internal energy. Therefore, the temperature decreases and the pressure decreases more strongly as the volume increases.

4.2.1   Example: Adiabatic Solar Corona

In Example 2.5.2 we have considered the case of an *isothermal* solar corona, in which $T \simeq 1.5 \times 10^6$ K, $n_0 \simeq 4 \times 10^8$ cm^{-3} and $r_0 \simeq R_\odot = 6.96 \times 10^{10}$ cm. Let us consider now an *adiabatic* corona, keeping the remaining hypotheses of the previous example. The hydrostatic equilibrium equation (2.79) and the equation of state (2.80) are the same, that is,

$$\frac{dP}{dr} = -\frac{G\, M_\odot\, n\, m_H}{r^2} \tag{4.27}$$

and

$$P = 2nkT . \tag{4.28}$$

From these relations we have now

$$\frac{dP}{dr} = 2kT\,\frac{dn}{dr} + 2nk\,\frac{dT}{dr} = -\frac{G\, M_\odot\, n\, m_H}{r^2} , \tag{4.29}$$

which can be compared to (2.81). For the adiabatic process we can use Eq. (4.25), recalling that $\rho = 2n\, m_H$, obtaining

$$T = \frac{T_0}{n_0^{\gamma-1}}\, n^{\gamma-1} , \tag{4.30}$$

where the temperature T and the density n are related to the reference values T_0 and n_0. Differentiating (4.30) relative to r and substituting in (4.29) and simplifying, we get

$$\frac{2\gamma k T_0}{n_0^{\gamma-1}}\, n^{\gamma-1}\,\frac{dn}{dr} = -\frac{G\, M_\odot\, n\, m_H}{r^2} . \tag{4.31}$$

Using the same definition of the scale height h given by (2.83) with $T = T_0$, that is, $h = 2k\, T_0\, r_0^2 / G\, M_\odot\, m_H$, Eq. (4.31) can be written as

$$n^{\gamma-2}\, dn = -\frac{n_0^{\gamma-1}\, r_0^2}{\gamma h}\, r^{-2}\, dr , \tag{4.32}$$

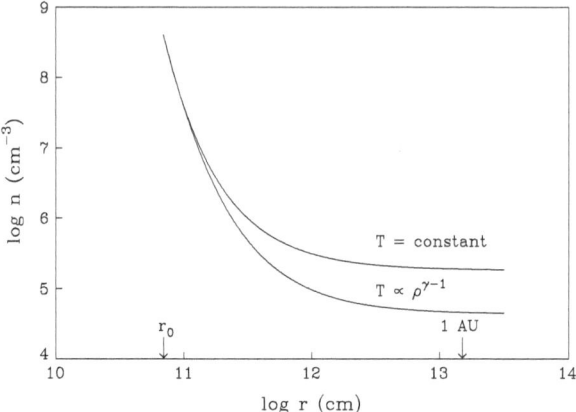

Fig. 4.1 Variation of the density in the solar corona according to the isothermal and adiabatic models

which can be integrated from r_0 to r, where the density varies from n_0 to n, resulting

$$n(r) = n_0 \left[1 - \frac{\gamma - 1}{\gamma} \frac{r_0^2}{h} \left(\frac{1}{r_0} - \frac{1}{r} \right) \right]^{\frac{1}{\gamma - 1}} . \tag{4.33}$$

The interesting solutions are those in which $(\gamma - 1)/\gamma < h/r_0 = 9.05 \times 10^9/6.96 \times 10^{10} \simeq 0.13$ according to the data of Example 2.5.2, so that we must have $\gamma < 1.15$. For example, taking $\gamma = 1.05$, we get from (4.33) the density $n \simeq 4.6 \times 10^4$ cm^{-3} at $r = 1$ AU $= 1.5 \times 10^{13}$ cm, which is 40 times smaller than the density obtained with the isothermal model. From (4.30), the coronal temperature in this region is $T = 9.5 \times 10^5$ K. Figure 4.1 shows the density variations in both cases, namely, the isothermal model of Example 2.5.2 and the adiabatic model with $\gamma = 1.05$.

4.2.2   Example: Entropy Flux Density

For a isentropic process we have

$$\frac{Ds}{Dt} = \frac{\partial s}{\partial t} + \mathbf{v} \cdot \nabla s = 0 . \tag{4.34}$$

Using the continuity equation in the form (1.8) and simplifying, we can write

$$\frac{\partial (\rho s)}{\partial t} + \nabla \cdot (\rho s \, \mathbf{v}) = 0 . \tag{4.35}$$

This is the entropy continuity equation, similar to the mass continuity equation (1.6). The product $\rho s \mathbf{v}$ (erg K^{-1} cm^{-2} s^{-1}) represents the *entropy flux density*. We should note that (4.35) may be written for any variable ψ that satisfies condition (4.34).

4.2.3 Example: Euler Equation in Isentropic Motion

Let us examine some consequences on the Euler equation in the case of an isentropic flow. From the first law of thermodynamics (see Eq. 4.3) we have

$$dq = T \, ds = de + P \, dV^* , \tag{4.36}$$

where dq (erg/g) is the amount of absorbed heat per unit mass in an infinitesimal process, de (erg/g) is the variation of the internal energy per unit mass, $P \, dV^*$ (erg/g) is the work done by the system per unit mass, $V^* = 1/\rho$ is the specific volume, and ds (erg K^{-1} g^{-1}) is the entropy variation in the process. Introducing the specific enthalpy,

$$h = e + P \, V^* = e + \frac{P}{\rho} , \tag{4.37}$$

we have

$$dh = de + P \, dV^* + V^* \, dP = T \, ds + V^* \, dP . \tag{4.38}$$

For an isentropic process, $ds = 0$ and $dh = V^* \, dP = dP/\rho$, so that

$$\nabla h = \frac{1}{\rho} \nabla P . \tag{4.39}$$

Substituting in Eq. (2.13), we obtain an expression of the Euler equation as

$$\frac{\partial \mathbf{v}}{\partial t} + (\mathbf{v} \cdot \nabla)\mathbf{v} = -\nabla h . \tag{4.40}$$

The isentropic, or adiabatic flow, as characterized by Eq. (4.2), is an example of a *barotropic* fluid, in which the equation of state can be written essentially in terms of the pressure P and density ρ. In more realistic situations, the variations of the pressure involve the gas temperature, and several complex physical processes must be taken into account.

4.3   Energy Conservation in a Fluid

Let us initially assume that the fluid is in a gravitational field, so that the gravitational force per unit volume can be written as

$$\mathbf{F} = \rho\,\mathbf{g} = -\rho\,\nabla\phi \;, \tag{4.41}$$

where ϕ is the gravitational potential of the field (units: $cm^2/s^2 = dyn\,cm\,g^{-1} = erg/g$), that is, the units of ϕ are energy/mass and the units of the product $\rho\phi$ are energy/volume. In this case, from (2.13) and (2.14) the Euler equation in the Lagrangian form can be written as

$$\frac{D\mathbf{v}}{Dt} = -\frac{1}{\rho}\,\nabla P + \mathbf{g} = -\frac{1}{\rho}\,(\nabla P + \rho\,\nabla\phi) \;. \tag{4.42}$$

Taking the scalar product with $\mathbf{v}$, we get

$$\frac{1}{2}\frac{Dv^2}{Dt} = -\frac{1}{\rho}\,\mathbf{v}\cdot(\nabla P + \rho\,\nabla\phi) \;. \tag{4.43}$$

Assuming that potential ϕ is constant in time, we have

$$\frac{D\phi}{Dt} = \frac{\partial\phi}{\partial t} + (\mathbf{v}\cdot\nabla)\,\phi = (\mathbf{v}\cdot\nabla)\,\phi \;. \tag{4.44}$$

Substituting (4.44) in (4.43) and rearranging terms, we get

$$\frac{D}{Dt}\left(\frac{1}{2}\rho v^2 + \rho\phi\right) = -(\mathbf{v}\cdot\nabla)\,P \;. \tag{4.45}$$

The total kinetic energy of the fluid is given by

$$E_c = \int_V \frac{1}{2}\,\rho v^2\,dV \;, \tag{4.46}$$

where the integral is performed in the volume V occupied by the fluid. Analogously, the total potential energy of the fluid is

$$E_p = \int_V \rho\phi\,dV \;. \tag{4.47}$$

Integrating equation (4.45) in the volume V, using (4.46) and (4.47),

$$\frac{D}{Dt}(E_c + E_p) = -\int_V (\mathbf{v}\cdot\nabla)\,P\,dV \;. \tag{4.48}$$

The left member of (4.48) represents the time rate of the total fluid energy, including the kinetic energy and gravitational potential energy. The right member of (4.48) gives essentially the rate of change of the work done by the pressure forces. This can be clearly seen using Eq. (1.7), replacing ρ by P, and using the divergence theorem (1.3), in order to simplify the right member of (4.48),

$$-\int_V (\mathbf{v} \cdot \nabla)\, P\, dV = -\oint P\, \mathbf{v} \cdot \mathbf{n}\, dS + \int_V P\, \nabla \cdot \mathbf{v}\, dV . \qquad (4.49)$$

Using this relation, (4.48) can be written as

$$\frac{D}{Dt}(E_c + E_p) = -\oint P\, \mathbf{v} \cdot \mathbf{n}\, dS + \int_V P\, \nabla \cdot \mathbf{v}\, dV . \qquad (4.50)$$

For an incompressible fluid, $\nabla \cdot \mathbf{v} = 0$ (Eq. 1.9), so that

$$\frac{D}{Dt}(E_c + E_p) = -\oint P\, \mathbf{v} \cdot \mathbf{n}\, dS . \qquad (4.51)$$

In this case, there is a simpler interpretation: since the term $P\, dS$ is the force acting perpendicularly to the surface element dS, the integrand is essentially the time rate with which the pressure forces do work on the fluid that traverses the surface element, so that the integral gives the time variation of the total work done. Relations (4.48), (4.50) and (4.51) are alternative forms of the energy conservation equation of the fluid. In the following sections we will examine additional forms of this equation that are more adequate to the study of stellar winds and other astrophysical applications.

4.3.1 Example: Energy Deposition in the Interstellar Medium

The kinetic energy per unit mass of a stellar wind with terminal velocity v_f is of the order of $(1/2)v_f^2$. If $dM/dt = \dot{M}$ is the mass loss rate associated to the wind, from energy conservation the kinetic energy deposition in the interstellar medium per unit time is $(1/2)\,\dot{M}\, v_f^2$. Considering that the wind acts during a time interval Δt, the total kinetic energy transferred to the interstellar medium is $(1/2)\,\dot{M}\, v_f^2\, \Delta t$. For example, for a hot star with $v_f \simeq 2{,}000\,\mathrm{km/s}$ and $\dot{M} \sim 10^{-5}\, M_\odot/\mathrm{year}$, the transfer rate of kinetic energy is 1.2×10^{37} erg/s. Assuming that the corresponding evolution stage for massive stars may last about $\Delta t \sim 10^5$ year, the total kinetic energy due to a single star is 3.9×10^{49} erg. This value can be compared with the total energy associated to a Type II supernova explosion, which is of the order of 10^{49}–10^{51} erg.

In the case of cool giant stars, the deposition rate is lower, about 3.1×10^{31} erg/s, where we have adopted $\dot{M} \sim 10^{-6}\, M_\odot/\mathrm{year}$ and $v_f \simeq 10\,\mathrm{km/s}$. Taking

$\Delta t \simeq 10^6$ year for the duration of the red giant stage, the total kinetic energy transferred by the star is 9×10^{44} erg, which can be compared with the energy released in a nova explosion, which is of the order of 10^{43}–10^{44} erg. In this case, the affected region of the interstellar medium is considerably smaller than in the previous case, but this is counterbalanced by the larger number of low and intermediate mass stars relative to the hot, massive stars that originate supernovae.

4.4   The Energy Equation in the Lagrangian Form

In Sect. 4.2 we studied the energy equation in an adiabatic process. In the general case, the gas interacts with the surrounding regions and there is some energy exchange. From the first law of thermodynamics, Eq. (4.3), and using specific quantities q^*, e e V^* we can write

$$dq^* = de + P \, dV^* = de + P \, d\left(\frac{1}{\rho}\right). \tag{4.52}$$

Let us consider a Lagrangian fluid element. We will be interested in the energy exchange involving this element. From (4.52),

$$\begin{cases} \dfrac{De}{Dt} + P \, \dfrac{D(1/\rho)}{Dt} = \dfrac{1}{\rho} \, Q \\[4mm] \rho \left[\dfrac{De}{Dt} + P \, \dfrac{D(1/\rho)}{Dt} \right] = Q \, , \end{cases} \tag{4.53}$$

where Q is now the rate of energy generated by the external forces, with units $\mathrm{erg\,cm^{-3}\,s^{-1}}$; this can be a negative quantity, as for example in the case of radiative losses. In the general case, this term should include all possible energy exchange forms, the most important of which are:

• Chemical sources/sinks,
• Nuclear sources/sinks,
• Energy transferred by conduction,
• Energy transferred by radiation,
• Dissipated mechanic energy,
• Dissipated acoustic energy.

In the case of stellar winds, the four last items are generally important. Let $\mathbf{q}$ be the transferred *energy flux* ($\mathrm{erg\,cm^{-2}\,s^{-1}}$). The energy transferred per cubic centimeter per second is $\nabla \cdot \mathbf{q}$, so that Eq. (4.53) is written as

$$\rho \left[\frac{De}{Dt} + P \, \frac{D(1/\rho)}{Dt} \right] = -\nabla \cdot \mathbf{q} \, . \tag{4.54}$$

The minus sign (–) is due to the definition of **q**, as the energy *transferred* or *lost* by the gas. In a more detailed way we can define:

Q_A: Acoustic our mechanic energy deposition rate (units: $\mathrm{erg\,cm^{-3}\,s^{-1}}$),
Q_R: Radiative energy deposition rate ($\mathrm{erg\,cm^{-3}\,s^{-1}}$),
  $\mathbf{q}_c$: Energy flux transferred by conduction (units: $\mathrm{erg\,cm^{-2}\,s^{-1}}$). The corresponding energy is $\nabla \cdot \mathbf{q}_c$ ($\mathrm{erg\,cm^{-3}\,s^{-1}}$).

Considering the last four processes mentioned above, we have

$$\rho \left[\frac{De}{Dt} + P \, \frac{D(1/\rho)}{Dt} \right] = Q_A + Q_R - \nabla \cdot \mathbf{q}_c \ . \tag{4.55}$$

Usually, the conductive flux is written in terms of the thermal conductivity K, given in $\mathrm{erg\,cm^{-1}\,s^{-1}\,K^{-1}}$:

$$\mathbf{q}_c = -K \, \nabla T \ . \tag{4.56}$$

In terms of the total derivative, the continuity equation can be written as (see Eqs. 1.6 and 2.12):

$$\frac{D\rho}{Dt} + \rho \, \nabla \cdot \mathbf{v} = 0 \ . \tag{4.57}$$

Since

$$\frac{D(1/\rho)}{Dt} = -\frac{1}{\rho^2} \frac{D\rho}{Dt} \ ,$$

we have

$$\rho \, \frac{D(1/\rho)}{Dt} = -\frac{1}{\rho} \frac{D\rho}{Dt} = \nabla \cdot \mathbf{v} \ . \tag{4.58}$$

Substituting (4.58) in (4.54), we have

$$\rho \, \frac{De}{Dt} + P \, \nabla \cdot \mathbf{v} = -\nabla \cdot \mathbf{q} \ . \tag{4.59}$$

Alternatively, substituting (4.58) in (4.55), we get

$$\rho \, \frac{De}{Dt} + P \, \nabla \cdot \mathbf{v} = Q_A + Q_R - \nabla \cdot \mathbf{q}_c \ . \tag{4.60}$$

In Eqs. (4.59) or (4.60), the first term of the first member represents the internal energy variation, the second term represents the work done by the system, and the second member represents the absorbed "heat" by the processes of energy deposition, conduction, etc. On the other hand, the kinetic energy conservation equation may be obtained from the Euler equation (see Eq. 2.15):

$$\frac{D\mathbf{v}}{Dt} = \frac{\partial \mathbf{v}}{\partial t} + (\mathbf{v} \cdot \nabla)\mathbf{v} = -\frac{1}{\rho} \nabla P + \frac{1}{\rho} \mathbf{F} \,.$$

Taking the scalar product,

$$\mathbf{v} \cdot \frac{D\mathbf{v}}{Dt} = -\frac{1}{\rho} \mathbf{v} \cdot \nabla P + \frac{1}{\rho} (\mathbf{v} \cdot \mathbf{F}) \,,$$

$$\rho \frac{D(v^2/2)}{Dt} + \mathbf{v} \cdot \nabla P = \mathbf{v} \cdot \mathbf{F} \,, \tag{4.61}$$

which is the kinetic energy conservation equation. The first term of the first member represents the kinetic energy variation. The second term represents the work done by the pressure forces and the second member represents the work done by the external forces. Adding (4.59) and (4.61),

$$\rho \frac{D}{Dt}(v^2/2 + e) + P(\nabla \cdot \mathbf{v}) + \mathbf{v} \cdot \nabla P = \mathbf{v} \cdot \mathbf{F} - \nabla \cdot \mathbf{q}.$$

Since

$$\nabla \cdot (P\,\mathbf{v}) = P(\nabla \cdot \mathbf{v}) + \mathbf{v} \cdot \nabla P \,,$$

we have

$$\begin{cases} \rho \dfrac{D(v^2/2 + e)}{Dt} + \nabla \cdot (P\,\mathbf{v}) = \mathbf{v} \cdot \mathbf{F} - \nabla \cdot \mathbf{q} \\[2mm] \rho \dfrac{D(v^2/2 + e)}{Dt} + \nabla \cdot (P\,\mathbf{v}) = \mathbf{v} \cdot \mathbf{F} + Q_A + Q_R - \nabla \cdot \mathbf{q}_c \,, \end{cases} \tag{4.62}$$

where the second equation was obtained using (4.60). This equation is the *energy equation in Lagrangian form*. The first term of the first member represents the total energy variation, including kinetic and internal energy ($\mathrm{erg\,cm^{-3}\,s^{-1}}$). The second term represents the work done by the pressure forces. The first term of the second member represents the work done by the external forces, and the remaining terms represent the transferred energy, gained or lost by the gas, by the processes of conduction, etc.

4.5   The Energy Equation in the Eulerian Form

Considering a function ψ that is continuous and differentiable, we may write $D\psi/Dt = \partial\psi/\partial t + \mathbf{v} \cdot \nabla\psi$, so that

$$\rho \frac{D\psi}{Dt} = \rho \frac{\partial\psi}{\partial t} + \rho\,\mathbf{v}\cdot\nabla\psi = \left[\frac{\partial(\rho\psi)}{\partial t} - \psi \frac{\partial\rho}{\partial t}\right] + \left[\nabla\cdot(\rho\,\psi\,\mathbf{v}) - \psi\,\nabla\cdot(\rho\,\mathbf{v})\right], \quad (4.63)$$

where we have used the fact that

$$\nabla \cdot (\rho\,\psi\,\mathbf{v}) = \psi\,\nabla \cdot (\rho\,\mathbf{v}) + \rho\,\mathbf{v} \cdot \nabla\psi \ .$$

Using the continuity equation, (4.63) is written as

$$\rho \frac{D\psi}{Dt} = \frac{\partial(\rho\,\psi)}{\partial t} + \nabla \cdot (\rho\,\psi\,\mathbf{v}) \ . \qquad (4.64)$$

Using (4.64) and considering $\psi = v^2/2 + e$, we get from the first Eq. (4.62):

$$\frac{\partial[\rho(v^2/2 + e)]}{\partial t} + \nabla \cdot [\rho\,\mathbf{v}(v^2/2 + e + P/\rho)] = \mathbf{v} \cdot \mathbf{F} - \nabla \cdot \mathbf{q} \ , \qquad (4.65)$$

that can be written as

$$\begin{cases} \dfrac{\partial[\rho(v^2/2 + e)]}{\partial t} + \nabla \cdot [\rho\,\mathbf{v}(v^2/2 + e + P/\rho) + \mathbf{q}] = \mathbf{v} \cdot \mathbf{F} \\[4mm] \dfrac{\partial[\rho(v^2/2 + e)]}{\partial t} + \nabla \cdot [\rho\,\mathbf{v}(v^2/2 + e + P/\rho) + \mathbf{q}_c] = \mathbf{v} \cdot \mathbf{F} + Q_A + Q_R \ . \end{cases}$$
$$(4.66)$$

The second of these equations was obtained using the second Eq. (4.62). Analogously to (4.62), Eq. (4.66) expresses the fact that the time variation of the total energy of the gas ($\mathrm{erg\,cm^{-3}\,s^{-1}}$) results from an energy flux through a volume element in the gas, including the gas kinetic energy, the internal energy, the work done by the pressure forces, the energy transferred by conduction, etc., and the work done by the external forces. Relation (4.66) can be considered as the *Eulerian form* of the energy equation.

4.5.1   Example: Bernoulli Equation

Let us consider an application of the energy equation as given by the second Eq. (4.66) in the case of a spherically symmetric, unidimensional steady flow. Let us neglect the electron conduction and take parameter Q ($\mathrm{erg\,cm^{-3}\,s^{-1}}$) as the total

energy deposition rate by processes such as acoustic energy, etc. The external force
is the gravitational force of a star with mass M_*, given by

$$\mathbf{v} \cdot \mathbf{F} = \rho \, \mathbf{v} \cdot \mathbf{g} = -\rho \, v \, \frac{G \, M_*}{r^2} \, . \tag{4.67}$$

In this case, from (4.66) we have

$$\frac{1}{r^2} \frac{d}{dr} \left[r^2 \rho \, v \, (v^2/2 + e + P/\rho) \right] = -\rho \, v \, \frac{G \, M_*}{r^2} + Q \, ,$$

that can be written as

$$\frac{d}{dr} \left(\frac{v^2}{2} + e + \frac{P}{\rho} - \frac{G \, M_*}{r} \right) = \frac{4 \, \pi \, r^2 \, Q}{\dot{M}} \, , \tag{4.68}$$

where we have used the fact that $r^2 \rho \, v = \dot{M}/4 \, \pi = $ constant in the case of a steady
flow. Since Q is defined as the energy transfer rate per unit volume (erg cm^{-3} s^{-1}),
the term $4 \, \pi \, r^2 \, Q/\dot{M} = d\epsilon/dr$ represents the transferred energy gradient per unit
mass (units: erg g^{-1} cm^{-1}). Including an additional external force F (dyn/cm^3), it
is easy to see that an additional term given by F/ρ should be added to the second
member of (4.68). The energy equation in this form is called the *Bernoulli equation*.

4.6   Fluid in a Conservative Force Field

Let us consider a fluid in a conservative field force, where $\mathbf{F} = \rho \, \mathbf{g} = -\rho \, \nabla \phi$, and
ϕ is the force field potential (Eq. 4.41). In this case we have

$$\mathbf{v} \cdot \mathbf{F} = -\rho \, \mathbf{v} \cdot \nabla \phi \, . \tag{4.69}$$

From (4.63), with $\psi = \phi$, we have

$$\rho \, \frac{D\phi}{Dt} = \rho \, \frac{\partial \phi}{\partial t} + \rho \, \mathbf{v} \cdot \nabla \phi \, . \tag{4.70}$$

Adding (4.70) and (4.62) and substituting (4.69),

$$\frac{\rho \, D(v^2/2 + e + \phi)}{Dt} + \nabla \cdot (P \, \mathbf{v}) = -\nabla \cdot \mathbf{q} + \rho \, \frac{\partial \phi}{\partial t} \, .$$

Using again (4.64), with $\psi = v^2/2 + e + \phi$, we get

$$\frac{\partial [\rho(v^2/2 + e + \phi)]}{\partial t} + \nabla \cdot [\rho \, \mathbf{v}(v^2/2 + e + \phi + P/\rho) + \mathbf{q}] = 0 \tag{4.71}$$

where we have assumed that the potential does not change with time, that is, $\partial \phi / \partial t = 0$.

4.7   The Energy Flux

Let us integrate (4.71) in a volume V. We get

$$\frac{\partial}{\partial t} \int_V \rho \left(\frac{v^2}{2} + e + \phi \right) dV = - \int_V \nabla \cdot \left[\rho \, \mathbf{v} \left(\frac{v^2}{2} + e + \phi + \frac{P}{\rho} \right) + \mathbf{q} \right] dV . \quad (4.72)$$

Using the divergence theorem, (4.72) is written as

$$\frac{\partial}{\partial t} \int_V \rho \left(\frac{v^2}{2} + e + \phi \right) dV = - \oint \left[\rho \, \mathbf{v} \left(\frac{v^2}{2} + e + \phi + \frac{P}{\rho} \right) + \mathbf{q} \right] \cdot \mathbf{n} \, dS . \quad (4.73)$$

This is analogous to Eq. (1.2): The left member of (1.2) gives the time rate of change of the *mass* inside volume V, which led us to call the product $\rho \, \mathbf{v}$ as the *mass flux vector*. Analogously, in (4.73) the left member gives the time rate of change of the gas *total energy* in the considered volume, which leads us to define the *energy flux ϵ* as

$$\epsilon = \rho \, \mathbf{v} \left(\frac{v^2}{2} + e + \phi + \frac{P}{\rho} \right) + \mathbf{q} , \quad (4.74)$$

(units: $\mathrm{erg \, cm}^{-2} \mathrm{s}^{-1}$). We notice that the total energy is a *scalar* quantity, and the energy flux is a *vector*. This flux corresponds to the total energy crossing a unit surface element perpendicularly to the gas velocity per unit time. The components of the integral in the second member of (4.73) are:

$\oint \rho \mathbf{v} \, (v^2/2) \cdot \mathbf{n} \, dS \rightarrow$ transferred kinetic energy per unit time,
$\oint \rho \mathbf{v} \, e \cdot \mathbf{n} \, dS \rightarrow$ transferred internal energy per unit time,
$\oint \rho \mathbf{v} \, \phi \cdot \mathbf{n} \, dS \rightarrow$ transferred potential energy per unit time,
$\oint P \, \mathbf{v} \cdot \mathbf{n} \, dS \rightarrow$ work done by the pressure forces per unit time,
$\oint \mathbf{q} \cdot \mathbf{n} \, dS \rightarrow$ transferred energy by conduction, etc., per unit time.

4.7.1   Example: Spherically Symmetric Conductive Flux

Let us initially consider the case where the flux $\mathbf{q}$ is the conductive flux, given by

$$\mathbf{q} = \mathbf{q}_c = -K \, \nabla T \quad (4.75)$$

(see Eq. 4.56). We assume a spherically symmetric, unidimensional flow, and the external force is the central star gravitational force. $\mathbf{v} \cdot \mathbf{F} = \rho \, \mathbf{v} \cdot \mathbf{g} = -\rho \, v \, (G \, M_* / r^2)$ (see Eq. 4.67). From the first Eq. (4.66) and using (1.15) we have,

$$\frac{\partial}{\partial t} \left[\rho \left(\frac{v^2}{2} + e \right) \right] + \frac{1}{r^2} \frac{\partial}{\partial r} \left[r^2 \, \rho \, v \left(\frac{v^2}{2} + e + \frac{P}{\rho} \right) - r^2 \, K \, \frac{\partial T}{\partial r} \right] = -\frac{G \, M_* \, \rho \, v}{r^2} \, . \tag{4.76}$$

For a stationary flow,

$$\frac{d}{dr} \left[r^2 \, \rho \, v \left(\frac{v^2}{2} + e + \frac{P}{\rho} \right) - r^2 \, K \, \frac{dT}{dr} \right] + (r^2 \, \rho \, v) \, \frac{G \, M_*}{r^2} = 0 \, . \tag{4.77}$$

In this case, form the continuity equation, $r^2 \, \rho \, v = $ constant, so that this equation can be integrated, resulting

$$(4 \, \pi \, r^2 \, \rho \, v) \left(\frac{v^2}{2} + e + \frac{P}{\rho} - \frac{G \, M_*}{r} \right) - 4 \, \pi \, r^2 \, K \, \frac{dT}{dr} = \text{constant} \, . \tag{4.78}$$

We see that in steady state, *the total energy flux is constant through a spherical surface.*

4.7.2 Example: Conductive Flux and Radiative Force

Let us now consider a spherically symmetric, unidimensional flow in steady state, including a conductive flux, and where the external forces are radiative and gravitational, characterized by the ratio (the Γ parameter)

$$\Gamma_r = \frac{g_r}{g_*} \tag{4.79}$$

(see Eq. 2.75). Analogously to Eq. (4.77), we may write from (4.66),

$$\frac{d}{dr} \left[r^2 \, \rho \, v \left(\frac{v^2}{2} + e + \frac{P}{\rho} \right) - r^2 \, K \, \frac{dT}{dr} \right] + (1 - \Gamma_r) \, (r^2 \, \rho \, v) \left(\frac{G \, M_*}{r^2} \right) = 0 \, . \tag{4.80}$$

If $\Gamma_r = 0$, (4.80) is reduced to (4.77). If Γ_r is constant, we have

$$\pi r^2 \, \rho \, v \left[\frac{v^2}{2} + e + \frac{P}{\rho} - \frac{GM_*}{r} \, (1 - \Gamma_r) \right] - 4 \pi r^2 \, K \frac{dT}{dr} = \text{constant} \, , \tag{4.81}$$

Which can be compared to (4.78).

Exercises

4.1. Show that, for an adiabatic expansion of an ideal, monatomic gas, we have $d \ln P / d \ln \rho = 5/3$.

4.2. Show that relations (4.19) and (4.20) are still valid for a perfect gas with molecular weight μ and characterized by the specific heats and gas constant per gram, if we replace the constant by $\mathcal{R}/\mu$.

4.3. Prove Eqs. (4.31)–(4.33).

4.4. Show that the energy flux ϵ defined by (4.74) can be written as

$$\epsilon = \rho\, \mathbf{v}\left(\frac{v^2}{2} + \phi + h \right) + \mathbf{q}\,,$$

where h is the specific enthalpy of the system.

4.5. Prove Eqs. (4.78) and (4.81).

Bibliography

Cassinelli, J.P.: Stellar winds. Ann. Rev. Astron. Astrophys. **17**, 275 (1979) (Excellent review article on stellar winds, with the application of the hydrodynamic equations, in particular the energy equation in stellar envelopes. See also Holzer, T. E., Axford, W. I., Ann. Rev. Astron. Astrophys. vol. 8, p. 31, 1970)

Lamers, H.J.G.L.M., Cassinelli, J.P.: Introduction to Stellar Winds. Cambridge University Press, Cambridge (1999) (Referred to in Chapter 1. Presents detailed discussions of the energy equation for stellar winds applied both to hot and cool stars. The appendix includes an overview of the main thermodynamic equations)

Mihalas, D.: Stellar Atmospheres. Freeman, San Francisco (1978) (Referred to in Chapter 1. Discusses the fluid equations applied to stellar winds, in particular the energy equation)

Reif, F.: Fundamentals of Statistical and Thermal Physics. McGraw-Hill, New York (1965) (Referred to in Chapter 3. Includes a good discussion of adiabatic processes)

Chapter 5
Cartesian Tensor Notation

Abstract This chapter presents the Cartesian tensor notation, or index notation, and shows how the main vector operations can be written in this notation.

5.1 Introduction

In the general form, the hydrodynamic equations can be written in a relatively simple way using the *Cartesian tensor notation*, instead of the vector notation we have used in the previous chapters. The tensor notation is used in some texts on stellar winds, especially when a detailed treatment is given of the stellar radiation transfer. Let us review the main characteristics of this notation.

5.2 Vectors

A vector can be characterized by its magnitude and direction, or by three Cartesian components,

$$\mathbf{v} = \begin{pmatrix} v_x \\ v_y \\ v_z \end{pmatrix} \tag{5.1}$$

which is the same as

$$\mathbf{v} = v_x \, \mathbf{i} + v_y \, \mathbf{j} + v_z \, \mathbf{k} \,, \tag{5.2}$$

where $\mathbf{i}, \mathbf{j}, \mathbf{k}$ are the unit vectors of the axes x, y, z, respectively. According to the *Cartesian tensor notation* or *index notation*, (5.1) can be written as

$$\mathbf{v} = v_i \,. \tag{5.3}$$

W.J. Maciel, *Hydrodynamics and Stellar Winds: An Introduction*, Undergraduate Lecture Notes in Physics, DOI 10.1007/978-3-319-04328-9__5,
© Springer International Publishing Switzerland 2014

It is implicitly understood that i varies from 1 to 3, and that $v_1 = v_x$, $v_2 = v_y$ and $v_3 = v_z$.

The coordinates of vector $\mathbf{v}$ in a new coordinate system x', y', z' obtained by the rotation of the original axes x, y, z can be written as

$$\mathbf{v}' = \begin{pmatrix} v'_x \\ v'_y \\ v'_z \end{pmatrix} = v'_i \ . \tag{5.4}$$

We know that $\mathbf{v}'$ can be obtained from $\mathbf{v}$ by

$$\mathbf{v}' = M \, \mathbf{v} \ , \tag{5.5}$$

where M is matrix that describes the coordinate transformation, or rotation matrix. Calling a_{ij} the elements of M, Eq. (5.5) can be written as

$$\begin{pmatrix} v'_x \\ v'_y \\ v'_z \end{pmatrix} = \begin{pmatrix} a_{11} \ a_{12} \ a_{13} \\ a_{21} \ a_{22} \ a_{23} \\ a_{31} \ a_{32} \ a_{33} \end{pmatrix} \begin{pmatrix} v_x \\ v_y \\ v_z \end{pmatrix} \ , \tag{5.6}$$

that is,

$$v'_i = \sum_{j=1}^{3} a_{ij} \, v_j \ . \tag{5.7}$$

The elements of the rotation matrix are the direction cosines of the directions of the new coordinate axes x', y', z' relative to axes $x, y, z . a_{ij}$ is the cosine of the angle between the new i axis ($i = 1, 2, 3; 1 = x', 2 = y', 3 = z'$) and the old j axis ($j = 1, 2, 3; 1 = x, 2 = y, 3 = z$).

5.2.1 Example: Rotation Around the z Axis

As an example, if the new system is obtained by rotation of the axes x e y by an angle θ around the z axis (Fig. 5.1), matrix M is

$$M = \begin{pmatrix} \cos\theta \ \ \sin\theta \ \ 0 \\ -\sin\theta \ \ \cos\theta \ \ 0 \\ 0 \ \ \ \ 0 \ \ \ 1 \end{pmatrix} \ . \tag{5.8}$$

Fig. 5.1 Rotation of axes
x and y around the z axis

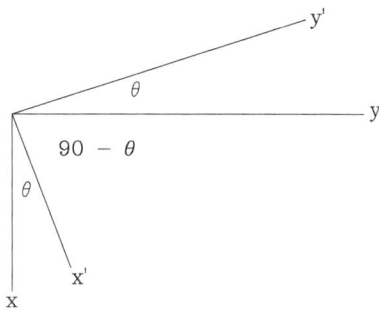

In this case, (5.6) becomes

$$
\begin{pmatrix} v'_x \\ v'_y \\ v'_z \end{pmatrix} = \begin{pmatrix} \cos\theta & \sin\theta & 0 \\ -\sin\theta & \cos\theta & 0 \\ 0 & 0 & 1 \end{pmatrix} \begin{pmatrix} v_x \\ v_y \\ v_z \end{pmatrix} , \tag{5.9}
$$

which can be written as

$$
\begin{cases} v'_x = \cos\theta\ v_x + \sin\theta\ v_y \\ v'_y = -\sin\theta\ v_x + \cos\theta\ v_y \\ v'_z = v_z . \end{cases} \tag{5.10}
$$

5.2.2   Example: Rotation Around the x Axis

We can use the same procedure of the previous example to obtain the rotation matrix in the case of a rotation around the x axis. Let us consider a vector **A** with coordinates $A_i = (1, 1, 0)$, that is, $A_x = 1$, $A_y = 1$, and $A_z = 0$. Considering a new coordinate system obtained by the counterclockwise rotation of the original system by an angle $\theta = 30°$ around the x axis, we can write

$$
A'_i = M\ A_i
$$

and the rotation matrix is

$$
M = \begin{pmatrix} 1 & 0 & 0 \\ 0 & \sqrt{3}/2 & 1/2 \\ 0 & -1/2 & \sqrt{3}/2 \end{pmatrix} .
$$

so that

$$A_i' = \begin{pmatrix} 1 \\ \sqrt{3}/2 \\ -1/2 \end{pmatrix} .$$

which can also be written as $A_x' = 1$, $A_y' = \sqrt{3/2}$, and $A_z' = -1/2$,
Using the *summation convention*, that is, the repetition of an index implies a summation over the possible values of the index, which in this case is from 1 to 3, we have from (5.7):

$$v_i' = a_{ij} \, v_j \, , \tag{5.11}$$

which represents the three equations for the components $v_1' = v_x'$, $v_2' = v_y'$, $v_3' = v_z'$:

$$\begin{cases} v_1' = a_{11} \, v_1 + a_{12} \, v_2 + a_{13} \, v_3 \\ v_2' = a_{21} \, v_1 + a_{22} \, v_2 + a_{23} \, v_3 \\ v_3' = a_{31} \, v_1 + a_{32} \, v_2 + a_{33} \, v_3 \, . \end{cases} \tag{5.12}$$

We have initially considered a vector as a quantity specified by its magnitude and direction. We can also define a vector in physical space as a quantity characterized by three numbers that transforms according to Eq. (5.11) in a change of coordinate systems.

5.3   Cartesian Tensors

Apart from the summation convention, we have used in (5.11) the "dimension convention", that is, we have assumed that the vector has three components, which are representative of the three dimensions of the physical space. In a r-dimensional space, the number of components of a vector is $r^1 = r$. As an extension of the vector concept, a *tensor of order n* in a r-dimensional space has r^n components. For example, for $r = 3$ (tridimensional physical space) the number of components of a tensor of order n is shown in Table 5.1

Table 5.1 Components of a tensor of order 3 in tridimensional space

n	Components	Notation	Name
0	$3^0 = 1$	A	Scalar
1	$3^1 = 3$	A_i	Vector
2	$3^2 = 9$	A_{ij}	Tensor of order 2
3	$3^3 = 27$	A_{ijk}	Tensor of order 3

Equation (5.11) for the vector transformation can be generalized for a Cartesian tensor of order n. In this case, considering a rotation of the coordinate system we have

$$A'_{rst...} = a_{ri}\, a_{sj}\, a_{tk} \ldots A_{ijk...}\,, \qquad (5.13)$$

where

$A_{ijk...}$ are the components of a Cartesian tensor of order n in the original system;
$A'_{rst...}$ are the components of the same tensor in the new coordinate system;
a_{ij} are the direction cosine between the new coordinate axes and the original axes.

For a second order tensor we can write (5.13) as

$$A'_{rs} = a_{ri}\, b_{sj}\, A_{ij} \qquad (5.14)$$

in matrix notation,

$$\begin{pmatrix} A'_{11} & A'_{12} & A'_{13} \\ A'_{21} & A'_{22} & A'_{23} \\ A'_{31} & A'_{32} & A'_{33} \end{pmatrix} = \begin{pmatrix} a_{11} & a_{12} & a_{13} \\ a_{21} & a_{22} & a_{23} \\ a_{31} & a_{32} & a_{33} \end{pmatrix} \begin{pmatrix} b_{11} & b_{12} & b_{13} \\ b_{21} & b_{22} & b_{23} \\ b_{31} & b_{32} & b_{33} \end{pmatrix} \begin{pmatrix} A_{11} & A_{12} & A_{13} \\ A_{21} & A_{22} & A_{23} \\ A_{31} & A_{32} & A_{33} \end{pmatrix} \qquad (5.15)$$

which represents nine algebraic equations:

$$A'_{11} = (a_{11}\, b_{11} + a_{12}\, b_{21} + a_{13}\, b_{31})\, A_{11} + (a_{11}\, b_{12} + a_{12}\, b_{22} + a_{13}\, b_{32})\, A_{21} + (a_{11}\, b_{13} + a_{12}\, b_{23} + a_{13}\, b_{33})\, A_{31}$$

$$A'_{12} = (a_{11}\, b_{11} + a_{12}\, b_{21} + a_{13}\, b_{31})\, A_{12} + (a_{11}\, b_{12} + a_{12}\, b_{22} + a_{13}\, b_{32})\, A_{22} + (a_{11}\, b_{13} + a_{12}\, b_{23} + a_{13}\, b_{33})\, A_{32}$$

$$A'_{13} = (a_{11}\, b_{11} + a_{12}\, b_{21} + a_{13}\, b_{31})\, A_{13} + (a_{11}\, b_{12} + a_{12}\, b_{22} + a_{13}\, b_{32})\, A_{23} + (a_{11}\, b_{13} + a_{12}\, b_{23} + a_{13}\, b_{33})\, A_{33}$$

$$A'_{21} = (a_{21}\, b_{11} + a_{22}\, b_{21} + a_{23}\, b_{31})\, A_{11} + (a_{21}\, b_{12} + a_{22}\, b_{22} + a_{23}\, b_{32})\, A_{21} + (a_{21}\, b_{13} + a_{22}\, b_{23} + a_{23}\, b_{33})\, A_{31}$$

$$A'_{22} = (a_{21}\, b_{11} + a_{22}\, b_{21} + a_{23}\, b_{31})\, A_{12} + (a_{21}\, b_{12} + a_{22}\, b_{22} + a_{23}\, b_{32})\, A_{22} + (a_{21}\, b_{13} + a_{22}\, b_{23} + a_{23}\, b_{33})\, A_{32}$$

$$A'_{23} = (a_{21}\, b_{11} + a_{22}\, b_{21} + a_{23}\, b_{31})\, A_{13} + (a_{21}\, b_{12} + a_{22}\, b_{22} + a_{23}\, b_{32})\, A_{23} + (a_{21}\, b_{13} + a_{22}\, b_{23} + a_{23}\, b_{33})\, A_{33}$$

$$A'_{31} = (a_{31}\, b_{11} + a_{32}\, b_{21} + a_{33}\, b_{31})\, A_{11} + (a_{31}\, b_{12} + a_{32}\, b_{22} + a_{33}\, b_{32})\, A_{21} + (a_{31}\, b_{13} + a_{32}\, b_{23} + a_{33}\, b_{33})\, A_{31}$$

$$A'_{32} = (a_{31} b_{11} + a_{32} b_{21} + a_{33} b_{31}) A_{12} + (a_{31} b_{12} + a_{32} b_{22} +$$
$$a_{33} b_{32}) A_{22} + (a_{31} b_{13} + a_{32} b_{23} + a_{33} b_{33}) A_{32}$$
$$A'_{33} = (a_{31} b_{11} + a_{32} b_{21} + a_{33} b_{31}) A_{13} + (a_{31} b_{12} + a_{32} b_{22} +$$
$$a_{33} b_{32}) A_{23} + (a_{31} b_{13} + a_{32} b_{23} + a_{33} b_{33}) A_{33}.$$

Comparison of the tensor relation (5.14) with the matrix notation (5.15) or with the last nine equations clearly shows the conciseness and elegance of the former.

5.4   Vector Operations

The Cartesian tensor notation, used for example in Eqs. (5.3) and (5.11), allows more clarity and conciseness in most vector operations. Let us consider the most usual operations in this notation.

5.4.1   Example: Scalar Product

In vector notation we have

$$\mathbf{A} \cdot \mathbf{B} = A_x B_x + A_y B_y + A_z B_z \tag{5.16}$$

and using the index notation,

$$\mathbf{A} \cdot \mathbf{B} = A_i B_i . \tag{5.17}$$

Introducing the Kronecker delta function (which is a tensor of order 2),

$$\delta_{ij} = \begin{cases} 1, & \text{if } i = j \\ 0, & \text{if } i \neq j , \end{cases} \tag{5.18}$$

We may write

$$\mathbf{A} \cdot \mathbf{B} = A_i B_j \delta_{ij} . \tag{5.19}$$

5.4.2   Example: Gradient

For a scalar function f that is univocal, continuous and with continuous derivatives in a certain domain, the gradient is a function given by

$$\nabla f = \frac{\partial f}{\partial x} \mathbf{i} + \frac{\partial f}{\partial y} \mathbf{j} + \frac{\partial f}{\partial z} \mathbf{k} , \tag{5.20}$$

which can be written as

$$(\nabla f)_i = \frac{\partial f}{\partial x_i} \ . \tag{5.21}$$

5.4.3   Example: Divergence

The divergence of a vector or vector function **A**, that is also univocal, continuous and with continuous derivatives in a certain domain, is

$$\nabla \cdot \mathbf{A} = \frac{\partial A_x}{\partial x} + \frac{\partial A_y}{\partial y} + \frac{\partial A_z}{\partial z} \ , \tag{5.22}$$

which can be written as

$$\nabla \cdot \mathbf{A} = \frac{\partial A_i}{\partial x_i} \ . \tag{5.23}$$

5.4.4   Example: Vector Product

The vector product of **A** and **B** is written as

$$\mathbf{A} \times \mathbf{B} = \begin{vmatrix} \mathbf{i} & \mathbf{j} & \mathbf{k} \\ A_x & A_y & A_z \\ B_x & B_y & B_z \end{vmatrix}$$

$$\mathbf{A} \times \mathbf{B} = (A_y\,B_z - A_z\,B_y)\,\mathbf{i} + (A_z\,B_x - A_x\,B_z)\,\mathbf{j} + (A_x\,B_y - A_y\,B_x)\,\mathbf{k} \ , \tag{5.24}$$

which can be written as

$$(\mathbf{A} \times \mathbf{B})_i = \varepsilon_{ijk}\,A_j\,B_k \ , \tag{5.25}$$

where we have introduced the *permutation symbol* ε_{ijk}, also called Levi-Civita's antisymmetric symbol,

$$\varepsilon_{ijk} = \begin{cases} +1, & \text{if } i,j,k \text{ form a cyclical permutation:} \\ & 123, 231, 312 \\ -1, & \text{if } i,j,k \text{ form an anticyclical permutation:} \\ & 132, 213, 321 \\ 0 & \text{if } i,j,k \text{ do not form permutations,} \\ & i = j, \ i = k, \text{ or } j = k \ . \end{cases} \tag{5.26}$$

The permutation symbol is related to Kronecker's δ by the $\varepsilon - \delta$ *identity*:

$$\varepsilon_{ijk} \, \varepsilon_{irs} = \varepsilon_{jki} \, \varepsilon_{irs} = \delta_{jr} \, \delta_{ks} - \delta_{js} \, \delta_{kr} \ . \tag{5.27}$$

ε_{ijk} is a pseudotensor, since it does not transform exactly according to Eq. (5.13). The product $(\mathbf{A} \times \mathbf{B})_i$ is a pseudovector.

5.4.5   Example: Rotational

The rotational of vector $\mathbf{A}$ is

$$\nabla \times \mathbf{A} = \begin{vmatrix} \mathbf{i} & \mathbf{j} & \mathbf{k} \\ \partial/\partial x & \partial/\partial y & \partial/\partial z \\ A_x & A_y & A_z \end{vmatrix}$$

$$\nabla \times \mathbf{A} = \left(\frac{\partial A_z}{\partial y} - \frac{\partial A_y}{\partial z} \right) \mathbf{i} + \left(\frac{\partial A_x}{\partial z} - \frac{\partial A_z}{\partial x} \right) \mathbf{j} + \left(\frac{\partial A_y}{\partial x} - \frac{\partial A_x}{\partial y} \right) \mathbf{k} \ . \tag{5.28}$$

Analogously to Eq. (5.25) this equation can be written as

$$(\nabla \times \mathbf{A})_i = \varepsilon_{ijk} \, \frac{\partial}{\partial x_j} \, A_k \ , \tag{5.29}$$

5.4.6   Example: Stokes Theorem

In a univocal, continuous, vector field $\mathbf{A}$, with continuous derivatives in a domain R, the line integral of $\mathbf{A}$ along a closed continuous trajectory is related to the surface integral of R around the trajectory by the relation

$$\oint \mathbf{A} \cdot d\mathbf{x} = \oint (\nabla \times \mathbf{A}) \cdot \mathbf{n} \, dS \ , \tag{5.30}$$

which is *Stokes theorem*. In this expression, $\mathbf{n}$ is the unit vector of the surface element dS. Using the index notation, we have

$$\oint A_i \, dx_i = \oint \varepsilon_{ijk} \, \frac{\partial A_k}{\partial x_j} \, n_i \, dS \ . \tag{5.31}$$

5.4.7   Example: Gradient Theorem

Considering a scalar function A and calling V the volume contained by the closed area S, we have

$$\int_V \nabla A \, dV = \oint A \, \mathbf{n} \, dS \, . \tag{5.32}$$

in the index notation,

$$\int_V \frac{\partial A}{\partial x_i} \, dV = \oint A \, n_i \, dS \, . \tag{5.33}$$

5.4.8   Example: Divergence Theorem

The divergence theorem, also called Gauss' theorem, Ostrogradsky's formula, or Gauss-Ostrogradsky's theorem is

$$\int_V \nabla \cdot \mathbf{A} \, dV = \oint \mathbf{A} \cdot \mathbf{n} \, dS \, . \tag{5.34}$$

In terms of the index notation,

$$\int_V \frac{\partial A_i}{\partial x_i} \, dV = \oint A_i \, n_i \, dS \, . \tag{5.35}$$

5.4.9   Example: Laplacian

The Laplacian operator $(\nabla^2 A = \nabla \cdot \nabla A)$ in Cartesian coordinates is written as

$$\nabla^2 = \frac{\partial^2}{\partial x^2} + \frac{\partial^2}{\partial y^2} + \frac{\partial^2}{\partial z^2} \, . \tag{5.36}$$

Using tensor notation,

$$\nabla^2 = \frac{\partial^2}{\partial x_i \, \partial x_i} \tag{5.37}$$

so that

$$\nabla^2 A = \frac{\partial^2 A}{\partial x_i \, \partial x_i} \, . \tag{5.38}$$

5.4.10 Example: Total Derivative

For the total derivative, we may write

$$\frac{D\mathbf{A}}{Dt} = \frac{\partial \mathbf{A}}{\partial t} + (\mathbf{v} \cdot \nabla)\,\mathbf{A} \tag{5.39}$$

$$\frac{DA_i}{Dt} = \frac{\partial A_i}{\partial t} + v_k \, \frac{\partial A_i}{\partial x_k} \;. \tag{5.40}$$

5.5 The Divergence Theorem

The divergence theorem can be generalized to include tensors of any order. To do that, let us introduce the multiplication of Cartesian tensors, which can be divided into two classes, as follows.

5.5.1 Example: Outer Product

The *outer product* of two Cartesian tensors is the Cartesian tensor obtained by placing the original tensors side by side *without* using summation indexes. For example, the product of A_i and B_i is

$$A_i\,B_i = A_i\,B_j = C_{ij} \;, \tag{5.41}$$

that is, the outer product of two first order tensors is a second order tensor. For A_i and B_{ij},

$$A_i\,B_{ij} = A_i\,B_{jk} = C_{ijk} \;, \tag{5.42}$$

that is, the outer product of a second order tensor by a first order tensor is a third order tensor. For A_{ij} and B_{ij},

$$A_{ij}\,B_{ij} = A_{ij}\,B_{rs} = C_{ijrs} \;, \tag{5.43}$$

that is, the outer product of two second order tensors is a fourth order tensor.

Table 5.2 Inner and outer product

Inner product	Outer product
$1 \times 1 = 0$	$1 \times 1 = 2$
$1 \times 2 = 1$	$1 \times 2 = 3$
$2 \times 2 = 2$	$2 \times 2 = 4$
$\vdots$	$\vdots$
$n \times m = n + m - 2$	$n \times m = n + m$

5.5.2   Example: Inner Product

The process of identification of a pair of indexes of a tensor is called *contraction*. For example,

$A_{ijk} \rightarrow A_{iik}$ after contraction of indexes i and j;
$A_{ijk} \rightarrow A_{iji}$ after contraction of indexes i and k;
$A_{ijk} \rightarrow A_{ijj}$ after contraction of indexes j and k .

Since the summation convention is valid, the resulting tensor is a first order tensor. Generally, the contraction process reduces the order, or rank of a Cartesian tensor by a factor 2. We can then define the *inner product* of two Cartesian tensors as an outer product followed by a contraction involving the indexes of both tensors. For example, for tensors A_i and B_j, we get

$$A_i \, B_j \rightarrow A_i \, B_i \rightarrow C \ ,$$

that is, the inner product of two first order tensor is a scalar. Naturally, this is the well-known *scalar product*, or dot product, which is a particular case of the inner product. For tensors A_i e B_{ij},

$$A_i \, B_{ij} \rightarrow A_i \, B_{ij} \rightarrow C_i A_i \, B_{ij} \rightarrow A_i \, B_{ji} \rightarrow C_i \ ,$$

that is, the inner product of a second order tensor by a first order tensor is a first order tensor. For tensors A_{ij} e B_{ij},

$$A_{ij} \, B_{ij} \rightarrow A_{ij} \, B_{rs} \rightarrow A_{ij} \, B_{is} \rightarrow C_{ij}$$
$$A_{ij} \, B_{ij} \rightarrow A_{ij} \, B_{rs} \rightarrow A_{ij} \, B_{ri} \rightarrow C_{ij}$$
$$A_{ij} \, B_{ij} \rightarrow A_{ij} \, B_{rs} \rightarrow A_{ij} \, B_{js} \rightarrow C_{ij}$$
$$A_{ij} \, B_{ij} \rightarrow A_{ij} \, B_{rs} \rightarrow A_{ij} \, B_{rj} \rightarrow C_{ij} \ ,$$

that is, the inner product of two second order tensors is a second order tensor (see Table 5.2).

Considering the inner product of two Cartesian tensors, the divergence theorem (5.35) can be written in a generalized form

$$\int_V \frac{\partial A_{ijk...}}{\partial x_r} \, dV = \oint A_{ijk...} \, n_r \, dS \, .$$ (5.44)

For a second order tensor A_{ij},

$$\int_V \frac{\partial A_{ij}}{\partial x_r} \, dV = \oint A_{ij} \, n_r \, dS \, .$$ (5.45)

It should be noted that the integrand in the second member of (5.35) is the inner product of two first order tensors (two vectors), so that it is a *scalar* quantity. In Eq. (5.45), the integrand involves the inner product of a second order tensor by a first order tensor, so that the result is a *vector*.

Exercises

5.1. Determine the rotation matrix of a coordinate system characterized by the rotation of axes x and y by an angle θ around the z axis and an angle ϕ around the y axis.

5.2. Calculate the number of components of tensors of order 0, 1, 2, and 3 in a space having (a) four dimensions and (b) ten dimensions.

5.3. Expand the representation of a second order tensor given by the matrix notation (5.15) and prove the nine algebraic equations given.

5.4. Apply the definition of the permutation symbol ϵ_{ijk} given by (5.26) and prove Eq. (5.25).

5.5. (a) Given the function $f(x, y, z) = x^2 - 2xy + y^2 + 2z^2$, what are the coordinates of vector ∇f? (b) What is the divergence of a vector $\mathbf{v}$ characterized by the components $v_i = (2x, -2y + 2yz, -z^2)$?

Bibliography

Arfken, G.B. Weber, H.J., Harris, F.E.: Mathematical Methods for Physicists. Academic, New York (2012) (Referred to in Chapter 2. Includes the main equations in spherical and cylindrical coordinates)

Boas, M.L.: Mathematical Methods in the Physical Sciences. Wiley, New York (2005) (New edition of a popular text on mathematical methods of physics, with a good discussion of vectors and tensors)

Jeffreys, H., Jeffreys, B.: Methods of Mathematical Physics. Cambridge University Press, Cambridge (2000) (New edition of a classic book on mathematical methods of physics, with a good discussion of vector and tensor calculus)

McConnell, A.J.: Applications of Tensor Analysis. Dover, New York (2011) (Basic text on tensorial methods with applications to several fields, such as hydrodynamics and electricity)

Temple, G.: Cartesian Tensors: An Introduction. Dover, New York (2004) (Reedition of a 1960 introductory text on the theory of Cartesian tensors)

Trefil, J.S.: Introduction to the Physics of Fluids and Solids. Dover, New York (2010) (Referred to in Chapter 1. The appendix includes a discussion of Cartesian tensor notation)

Chapter 6
Fluid Equations in Tensor Form

Abstract This chapter presents the main fluid equations, namely the continuity, Euler and energy equations using the Cartesian tensor notation. The momentum flux tensor is defined in the framework of the Euler equation in the presence of external forces.

6.1 Introduction

Using the Cartesian tensor notation described in the previous chapter we can write the fluid equations – the continuity equation, Euler equation and the energy equation – in a concise and elegant form. The equations in this form are especially useful in the case of real fluids, where viscosity cannot be neglected, as we will see in Chap. 7. Moreover, in tensor form the equations favour some useful generalizations, particularly with the introduction of the momentum flux tensor, as we will see in Sect. 6.4. An interesting reference on the applications of the hydrodynamic equations in tensor form to astrophysical problems, including the continuity equation and Euler equation, is the book by Trefil (2010).

6.2 The Continuity Equation

We have seen in Chap. 1 that the continuity equation can be written as

$$\frac{\partial \rho}{\partial t} + \nabla \cdot (\rho \mathbf{v}) = 0 \,. \tag{6.1}$$

Using the Cartesian tensor notation, this equation becomes

$$\frac{\partial \rho}{\partial t} + \frac{\partial (\rho \, v_k)}{\partial x_k} = 0 \,. \tag{6.2}$$

W.J. Maciel, *Hydrodynamics and Stellar Winds: An Introduction*, Undergraduate
Lecture Notes in Physics, DOI 10.1007/978-3-319-04328-9_6,
© Springer International Publishing Switzerland 2014

Using the expression

$$\frac{\partial(\rho \, v_i \, v_k)}{\partial x_k} = \rho \, v_k \, \frac{\partial v_i}{\partial x_k} + v_i \, \frac{\partial(\rho \, v_k)}{\partial x_k} \, , \qquad (6.3)$$

Eq. (6.2) can be written as

$$\frac{\partial(\rho \, v_i \, v_k)}{\partial x_k} = \rho \, v_k \, \frac{\partial v_i}{\partial x_k} - v_i \, \frac{\partial \rho}{\partial t} \, . \qquad (6.4)$$

6.3   Euler Equation in the Absence of External Forces

In Chap. 2 we have seen that Euler equation in the case of a non-viscous fluid in the absence of external forces is

$$\frac{\partial \mathbf{v}}{\partial t} + (\mathbf{v} \cdot \nabla)\mathbf{v} = -\frac{1}{\rho} \, \nabla P \, . \qquad (6.5)$$

Using the Cartesian tensor notation, we get

$$\frac{\partial v_i}{\partial t} + v_j \, \frac{\partial v_i}{\partial x_j} = -\frac{1}{\rho} \, \frac{\partial P}{\partial x_i} \, . \qquad (6.6)$$

Using the expression

$$\frac{\partial(\rho \, v_i)}{\partial t} = \rho \, \frac{\partial v_i}{\partial t} + v_i \, \frac{\partial \rho}{\partial t} \, , \qquad (6.7)$$

Eq. (6.6) can be written as

$$\frac{\partial(\rho \, v_i)}{\partial t} = \left(-\frac{\partial P}{\partial x_i} - \rho \, v_j \, \frac{\partial v_i}{\partial x_j} \right) + v_i \, \frac{\partial \rho}{\partial t} = -\frac{\partial P}{\partial x_i} - \left(\rho \, v_j \, \frac{\partial v_i}{\partial x_j} - v_i \, \frac{\partial \rho}{\partial t} \right) . \qquad (6.8)$$

Substituting (6.4) in (6.8), we have

$$\frac{\partial(\rho \, v_i)}{\partial t} = -\left[\frac{\partial P}{\partial x_i} + \frac{\partial(\rho \, v_i \, v_k)}{\partial x_k} \right] . \qquad (6.9)$$

Equation (6.9) can be written in a simpler manner, defining the tensor Π_{ik} by

$$\Pi_{ik} = P \, \delta_{ik} + \rho \, v_i \, v_k \, . \qquad (6.10)$$

The units of Π_{ik} are dyn/cm^2 = (g cm s^{-2}) cm^{-2} = (g cm s^{-1}) cm^{-2} s^{-1}, that is, units of momentum per unit area per unit time. We can see that

$$\frac{\partial \Pi_{ik}}{\partial x_k} = \frac{\partial(P\,\delta_{ik} + \rho\,v_i\,v_k)}{\partial x_k} = \frac{\partial(P\,\delta_{ik})}{\partial x_k} + \frac{\partial(\rho\,v_i\,v_k)}{\partial x_k} = \frac{\partial P}{\partial x_i} + \frac{\partial(\rho\,v_i\,v_k)}{\partial x_k}. \quad (6.11)$$

Considering (6.9) and (6.11), we conclude that

$$\frac{\partial(\rho\,v_i)}{\partial t} = -\frac{\partial \Pi_{ik}}{\partial x_k}. \quad (6.12)$$

This is *Euler equation in the absence of external forces in Cartesian tensor notation.* This equation, which results from the combination of the continuity and motion equations, is equivalent to the momentum conservation equation written in a form similar to the continuity equation (6.2).

In terms of the total, or Lagrangian derivative, from Eqs. (6.10) and (6.12) we can write the equation of motion in the form

$$\frac{\partial(\rho\,v_i)}{\partial t} + \frac{\partial(\rho\,v_i\,v_k)}{\partial x_k} = -\frac{\partial(P\,\delta_{ik})}{\partial x_k}.$$

Using (6.6) or the continuity equation (6.2) and relation (5.40) with $A_i = v_i$, we have

$$\frac{\partial(\rho\,v_i)}{\partial t} + \frac{\partial(\rho\,v_i\,v_k)}{\partial x_k} = \rho\,\frac{\partial v_i}{\partial t} + \rho\,v_k\,\frac{\partial v_i}{\partial x_k} = \rho\,\frac{D v_i}{Dt}. \quad (6.13)$$

Therefore, from (6.13) and (6.12) we find

$$\rho\,\frac{D v_i}{Dt} = -\frac{\partial P}{\partial x_i}, \quad (6.14)$$

which obviously could have been obtained directly from Eq. (6.6).

6.4 The Momentum Flux Tensor

The tensor Π_{ik} defined by Eq. (6.10) is a symmetric, second order tensor, and its components can be written as

$$\Pi_{ik} = \begin{pmatrix} \Pi_{11} & \Pi_{12} & \Pi_{13} \\ \Pi_{21} & \Pi_{22} & \Pi_{23} \\ \Pi_{31} & \Pi_{32} & \Pi_{33} \end{pmatrix} \quad (6.15)$$

or

$$\Pi_{ik} = \begin{pmatrix} P + \rho\,v_x^2 & \rho\,v_x\,v_y & \rho\,v_x\,v_z \\ \rho\,v_x\,v_y & P + \rho\,v_y^2 & \rho\,v_y\,v_z \\ \rho\,v_x\,v_z & \rho\,v_y\,v_z & P + \rho\,v_z^2 \end{pmatrix} . \tag{6.16}$$

We see that Π_{ik} is a symmetric tensor. Let us show that Π_{ik} is associated to the fluid *momentum flux*. We know that the momentum associated with a volume element dV moving with velocity $\mathbf{v}$ is $\rho\,\mathbf{v}\,dV$ ($\mathrm{g\,cm^{-3}\,cm^3\,cm\,s^{-1}} = \mathrm{g\,cm\,s^{-1}}$). Therefore, the first member of Eq. (6.12) is simply the time rate of the fluid momentum per unit volume. Adding all volume elements in a volume V, we get

$$\int_V \frac{\partial(\rho\,v_i)}{\partial t}\,dV = \frac{\partial}{\partial t}\int_V \rho\,v_i\,dV = -\int_V \frac{\partial \Pi_{ik}}{\partial x_k}\,dV . \tag{6.17}$$

Using the divergence theorem in tensor form, Eq. (5.45),

$$\int_V \frac{\partial \Pi_{ik}}{\partial x_k}\,dV = \oint \Pi_{ik}\,n_k\,dS , \tag{6.18}$$

we get

$$\frac{\partial}{\partial t}\int_V \rho\,v_i\,dV = -\oint \Pi_{ik}\,n_k\,dS . \tag{6.19}$$

This equation is similar to Eq. (1.2),

$$\frac{\partial}{\partial t}\int_V \rho\,dV = -\oint \rho\,\mathbf{v}\cdot\mathbf{n}\,dS , \tag{6.20}$$

where the *mass flux* vector $\rho\,\mathbf{v} = \mathbf{j}$ ($\mathrm{g\,cm^{-2}\,s^{-1}}$) is associated to the time rate of the total mass contained in the considered volume. Equation (6.19) is also similar to Eq. (4.73), where the *energy flux* vector, given by $\boldsymbol{\epsilon} = \rho\,\mathbf{v}(v^2/2 + e + \phi + P/\rho) + \mathbf{q}$ ($\mathrm{erg\,cm^{-2}\,s^{-1}}$), is associated to the time rate of the total gas energy in the volume. Analogously, Eq. (6.19) shows that Π_{ik} is associated to the time rate of the momentum $\rho\,v_i\,dV$ of element dV, so that it is natural to call Π_{ik} the *momentum flux*, with units $\mathrm{dyn/cm^2} = \mathrm{g\,cm\,s^{-1}\,cm^{-2}\,s^{-1}}$, that is, units of momentum per $\mathrm{cm^2}$ and per second. In the present case, the momentum flux is a *second order tensor*, as the momentum is a *vector*, or *first order tensor*.

Finally, Eq. (6.12) shows that a change in the momentum in a given point in space is related to the momentum flux through a volume element around the considered point. The Euler equation in the form (6.12) is particularly useful in the case of viscous fluids, as we will see later.

6.5   Euler Equation with External Forces

To derive Eq. (6.12) we started with the Euler equation as in (6.5), which was obtained in the absence of external forces. Let us now consider the case where an external force $\mathbf{F}$ (dyn/cm^3) acts on the fluid. In this case, (6.5) can be written as

$$\frac{\partial \mathbf{v}}{\partial t} + (\mathbf{v} \cdot \nabla)\mathbf{v} = -\frac{1}{\rho} \nabla P + \frac{1}{\rho} \mathbf{F} . \tag{6.21}$$

Using tensor notation, (6.21) becomes

$$\frac{\partial v_i}{\partial t} + v_j \frac{\partial v_i}{\partial x_j} = -\frac{1}{\rho} \frac{\partial P}{\partial x_i} + \frac{1}{\rho} F_i . \tag{6.22}$$

Considering (6.5) and (6.22), we have in analogy with Eq. (6.8),

$$\frac{\partial (\rho\, v_i)}{\partial t} = \left(-\frac{\partial P}{\partial x_i} - \rho\, v_j \frac{\partial v_i}{\partial x_j} + F_i \right) + v_i \frac{\partial \rho}{\partial t}$$

$$\frac{\partial (\rho\, v_i)}{\partial t} = -\frac{\partial P}{\partial x_i} - \left(\rho\, v_j \frac{\partial v_i}{\partial x_j} - v_i \frac{\partial \rho}{\partial t} \right) + F_i , \tag{6.23}$$

Substituting (6.4) in (6.23),

$$\frac{\partial (\rho\, v_i)}{\partial t} = -\frac{\partial P}{\partial x_i} - \frac{\partial (\rho\, v_i\, v_k)}{\partial x_k} + F_i = -\left[\frac{\partial P}{\partial x_i} + \frac{\partial (\rho\, v_i\, v_k)}{\partial x_k} \right] + F_i , \tag{6.24}$$

which can be compared to (6.9). Considering (6.11), Eq. (6.24) can be written as

$$\frac{\partial (\rho\, v_i)}{\partial t} = -\frac{\partial \Pi_{ik}}{\partial x_k} + F_i , \tag{6.25}$$

This is the equivalent equation to (6.12) in the presence of external forces. It should be noted that Π_{ik} is a second order tensor, and the term $\partial \Pi_{ik}/\partial x_k$, which is the divergence of Π_{ik}, is a first order tensor (a vector), such as the force F_i.

6.6   The Energy Equation

We can also write the energy equation using the index notation. For example, using the Lagrangian form of this equation, as given by the first Eq. (4.62), we have

$$\rho \frac{D(v^2/2 + e)}{Dt} + \nabla \cdot (P\, \mathbf{v}) = \mathbf{v} \cdot \mathbf{F} - \nabla \cdot \mathbf{q} . \tag{6.26}$$

We have seen that this equation can be written in the Eulerian form, as in the first Eq. (4.66),

$$\frac{\partial[\rho(v^2/2 + e)]}{\partial t} + \nabla \cdot [\rho \, \mathbf{v}(v^2/2 + e + P/\rho) + \mathbf{q}] = \mathbf{v} \cdot \mathbf{F} \,. \tag{6.27}$$

Using the energy flux vector, this relation becomes

$$\frac{\partial[\rho(v^2/2 + e)]}{\partial t} + \nabla \cdot \boldsymbol{\epsilon} = \mathbf{v} \cdot \mathbf{F} \,. \tag{6.28}$$

With the index notation, we get

$$\frac{\partial[\rho(v^2/2 + e)]}{\partial t} + \frac{\partial \epsilon_k}{\partial x_k} = v_i \, F_i \,. \tag{6.29}$$

A similar procedure can be adopted for the remaining forms of the energy equation.

Exercises

6.1. Write the continuity equation in Cartesian tensor notation for an incompressible fluid.

6.2. Consider the transformation of a vector defined on the (x, y) plane. (a) Write the matrix that represents rotation of the coordinate axes by an angle θ. (b) Show that vector $\mathbf{v} = v_x \, \mathbf{i} + v_y \, \mathbf{j}$ follows the vector transformation rules, where v_x and v_y are the vector components along the axes x and y, respectively.

6.3. Show that the quantity Π_{ik} defined in Eq. (6.10) is a second order tensor that transforms according to condition (5.14). Consider a two-dimensional space (x, y), as in Exercise 6.2.

6.4. Consider the vector F_i and tensor Π_{ik} in a three-dimensional space. Show that (a) $F_i = \delta_{ik} \, F_k$, and (b) $\Pi_{i\ell} = \delta_{ik} \, \Pi_{k\ell}$.

6.5. In a gas containing different types of particles, we can define the momentum flux tensor Π_{ik}^s for a particle of type s, mass m_s and number density n_s. Assume that the gas is homogeneous with equipartition of energy, so that all different particles have the same isotropic velocity distribution, characterized by the same temperature, and show that the momentum flux for all types of particles is still given by Eq. (6.10).

Bibliography

Aris, R.: Vectors, Tensors, and the Basic Equations of Fluid Mechanics. Dover, New York (2012) (Recent printing of a 1962 book. Concise, intermediate level book, emphasizing the application of tensors to the equations of fluid mechanics)

Landau, L., Lifchitz, E.: Mécanique des Fluides. MIR, Moscou (1971) English edition: Fluid Mechanics. Butterworth-Heinemann, Boston (1995) (Referred to in Chapter 1. Presents a derivation of the momentum flux using Cartesian index notation)

Mihalas, D.: Stellar Atmospheres. Freeman, San Francisco (1978) (Referred to in Chapter 1. Complete advanced treatment of the fluid equations applied to stellar winds, the momentum flux tensor and radiation hydrodynamics. In particular, the tensor Π_{ik} is rigorously defined in terms of the velocity distributions of the different particles that constitute the gas)

Trefil, J.S.: Introduction to the Physics of Fluids and Solids. Dover, New York (2010) (Referred to in Chapter 1. Includes a discussion of the Euler and continuity equations in Cartesian tensor notation, apart from some exercises and applications)

Chapter 7
The Navier-Stokes Equation

Abstract The main goal of this chapter is to present the Navier-Stokes equation, both for incompressible and compressible fluids. The equation is written in the cartesian tensor notation and also in the usual vector form. The viscosity and rate of strain tensors are introduced, as well as the viscosity coefficients.

7.1 Viscous Fluids

Until now, we have considered the properties of *ideal fluids*, in which, in the absence of external forces, momentum transfer (and therefore forces acting in a volume unit) is due to the pressure gradient. The limitation of this treatment can be made evident by a simple example. Let us consider a layer in a fluid in equilibrium, above which another layer is moving with a constant velocity. In an ideal fluid, the motion would continue indefinitely, even if no external force is acting on the fluid. Obviously, in real fluids the upper layer would decelerate and eventually stop, so that the fluid would be again in equilibrium. The deceleration is caused by a *shear force* that acts as a friction force, and is a consequence of the fluid viscosity, which is a manifestation of irreversible processes of energy dissipation. The flow of a viscous fluid can be thought of as a pile of sheets moving relative to each other. In this case, the viscosity corresponds to the *friction* between any pair of sheets.

In the study of stellar winds only ideal fluids are generally considered, so that the developments of this chapter are not normally used. However, it is interesting to examine the modifications in the Euler equation in the case of viscous fluids, which may be useful in other applications. In this chapter we present an intuitive and simplified derivation of the Navier-Stokes equation for viscous fluids. More rigorous treatments can be found in the bibliography at the end of the chapter.

The continuity equation as obtained in Chap. 1 remains valid for viscous fluids, since it reflects mass conservation in the fluid. On the other hand, the Euler equation is considerably modified, both for compressible and incompressible fluids. Let us present the Euler equation again in tensor form, in the case of a fluid without external forces,

W.J. Maciel, *Hydrodynamics and Stellar Winds: An Introduction*, Undergraduate
Lecture Notes in Physics, DOI 10.1007/978-3-319-04328-9_7,
© Springer International Publishing Switzerland 2014

$$\frac{\partial(\rho\, v_i)}{\partial t} = -\frac{\partial \Pi_{ik}}{\partial x_k}\,. \tag{7.1}$$

We can consider the action of viscosity as a force $\mathscr{F}$ per unit volume (dyn/cm^3). In this case, (7.1) can be written as (see Eq. 6.25),

$$\frac{\partial(\rho\, v_i)}{\partial t} = -\frac{\partial \Pi_{ik}}{\partial x_k} + \mathscr{F}_i\,. \tag{7.2}$$

It is convenient to define the *viscosity tensor* σ_{ik} (dyn/cm^2), such that

$$\mathscr{F}_i = \frac{\partial \sigma_{ik}}{\partial x_k}\,. \tag{7.3}$$

Euler equation (7.2) becomes

$$\frac{\partial(\rho\, v_i)}{\partial t} = -\frac{\partial \Pi_{ik}}{\partial x_k} + \frac{\partial \sigma_{ik}}{\partial x_k} = -\frac{\partial}{\partial x_k}\,(\Pi_{ik} - \sigma_{ik})\,. \tag{7.4}$$

This equation can be compared with Eq. (6.12), where $\sigma_{ik} = 0$. Considering now a volume V in the fluid enclosed by a surface S, the total force acting on the fluid is

$$\int_V F_i(\text{total})\, dV = -\int_V \frac{\partial}{\partial x_k}\,(\Pi_{ik} - \sigma_{ik})\, dV = -\oint(\Pi_{ik} - \sigma_{ik})\, n_k\, dS\,,$$

that is,

$$\int_V \frac{\partial}{\partial x_k}\,(\Pi_{ik} - \sigma_{ik})\, dV = \oint(\Pi_{ik} - \sigma_{ik})\, n_k\, dS\,, \tag{7.5}$$

where we have used the divergence theorem in the form (5.45). Comparing (7.5) with (6.18) we see that, if $-\oint \Pi_{ik}\, n_k\, dS$ corresponds to the momentum variation by the pressure forces and by the motion of the fluid across the surface S, then $\oint \sigma_{ik}\, n_k\, dS$ is the corresponding term produced by the viscous forces acting on the fluid. In a microscopic description, this term corresponds to the momentum transferred through the surface S by inelastic collisions of the atoms close to the surface.

7.2   The Viscosity Tensor

The viscosity tensor σ_{ik} can be obtained in a general and rigorous way by an extension of elasticity theory, for example. Here we will present a much simpler, intuitive derivation, following the treatment given by Landau and Lifchitz (1971) and Trefil (2010).

Let us consider an isotropic fluid, that is, its properties do not depend on the considered direction. If the fluid is in equilibrium or in a uniform motion with velocity **v**, assumed the same in all points of the fluid, it is apparent that there will be no viscous forces. However, if there is some relative velocity among the different parts of the fluid, the viscous forces will be important. Therefore, we would expect that the tensor σ_{ik} is not proportional to the velocity v_i itself, but that it depends on the possible velocity variations, which can be measured by terms such as

$$\frac{\partial v_i}{\partial x_k}, \quad \frac{\partial^2 v_i}{\partial x_k \, \partial x_j}, \quad \cdots$$

Let us restrict our analysis to the linear terms in σ_{ik}, that is, neglecting higher order derivatives. The viscous forces would also not be important if the fluid is in uniform rotation, so that its velocity can be written as

$$\mathbf{v} = \boldsymbol{\omega} \times \mathbf{r}, \tag{7.6}$$

where $\boldsymbol{\omega}$ is the fluid angular velocity. Therefore, σ_{ik} must be zero if the velocity is constant and also if a relation such as (7.6) is satisfied. Recalling Eq. (5.25), we can write

$$\begin{cases} v_i = \varepsilon_{ijk} \, \omega_j \, x_k \\ v_k = \varepsilon_{kji} \, \omega_j \, x_i, \end{cases} \tag{7.7}$$

so that

$$\begin{cases} \dfrac{\partial v_i}{\partial x_k} = \varepsilon_{ijk} \, \omega_j \\[4mm] \dfrac{\partial v_k}{\partial x_i} = \varepsilon_{kji} \, \omega_j \,. \end{cases} \tag{7.8}$$

In order to satisfy both conditions, tensor σ_{ik} should be proportional to the terms

$$\frac{\partial v_i}{\partial x_k} + \frac{\partial v_k}{\partial x_i}$$

and

$$\delta_{ik} \frac{\partial v_r}{\partial x_r} \,.$$

In other words, we can write

$$\sigma_{ik} = \eta \left(\frac{\partial v_i}{\partial x_k} + \frac{\partial v_k}{\partial x_i} \right) + \lambda \, \delta_{ik} \frac{\partial v_r}{\partial x_r} \,, \tag{7.9}$$

where η and λ are the *dynamic viscosity coefficients*, which in general depend on the position in the fluid. The units of the coefficients are: $[\eta] = [\lambda] = \mathrm{dyn\,s\,cm}^{-2} = \mathrm{g\,cm}^{-1}\,\mathrm{s}^{-1}$. Some typical values of the viscosity coefficient η at room temperature are: $\eta \simeq 0.01\,\mathrm{g\,cm}^{-1}\,\mathrm{s}^{-1}$ (water) and $\eta \simeq 1.8 \times 10^{-4}\,\mathrm{g\,cm}^{-1}\,\mathrm{s}^{-1}$ (air).

Substituting (7.9) in (7.4),

$$\frac{\partial(\rho \, v_i)}{\partial t} = -\frac{\partial \Pi_{ik}}{\partial x_k} + \frac{\partial}{\partial x_k} \left[\eta \left(\frac{\partial v_i}{\partial x_k} + \frac{\partial v_k}{\partial x_i} \right) + \lambda \, \delta_{ik} \frac{\partial v_r}{\partial x_r} \right].$$

Using the definition of the tensor Π_{ik} given by (6.10),

$$\frac{\partial(\rho \, v_i)}{\partial t} = -\frac{\partial}{\partial x_k} \left[P_0 \, \delta_{ik} + \rho \, v_i \, v_k - \eta \left(\frac{\partial v_i}{\partial x_k} + \frac{\partial v_k}{\partial x_i} \right) - \lambda \, \delta_{ik} \frac{\partial v_r}{\partial x_r} \right], \tag{7.10}$$

where we have used the notation P_0 for the gas pressure. In terms of the total derivative, (7.10) can be written as (see Eq. 6.13),

$$\rho \frac{Dv_i}{Dt} = -\frac{\partial}{\partial x_k} \left[P_0 \, \delta_{ik} - \eta \left(\frac{\partial v_i}{\partial x_k} + \frac{\partial v_k}{\partial x_i} \right) - \lambda \, \delta_{ik} \frac{\partial v_r}{\partial x_r} \right] \tag{7.11}$$

Defining the tensor τ_{ik},

$$\tau_{ik} = \sigma_{ik} - P_0 \, \delta_{ik} = -P_0 \, \delta_{ik} + \eta \left(\frac{\partial v_i}{\partial x_k} + \frac{\partial v_k}{\partial x_i} \right) + \lambda \, \delta_{ik} \frac{\partial v_r}{\partial x_r} \,, \tag{7.12}$$

Eq. (7.11) becomes

$$\rho \frac{Dv_i}{Dt} = \frac{\partial \tau_{ik}}{\partial x_k} \,, \tag{7.13}$$

which can be compared with Eq. (6.14). We see that, if $\eta = \lambda = 0$, that is, if the viscosity is negligible, (7.11) or (7.13) are reduced to the previous case, given by Eq. (6.14), and P_0 is the pressure at any point of the fluid. In the present case, the concept of pressure is more complex, and the term $P_0 \, \delta_{ik}$ is the *static value of tensor* τ_{ik}, which is the so-called *stress tensor*. Tensor τ_{ik} is defined in such a way that $\tau_{ik} \, n_k$ is the force acting on a surface element of unit k. Therefore, τ_{ii} is the normal component acting on an element with unit n_i. Introducing the tensors

$$e_{ik} = \frac{1}{2} \left(\frac{\partial v_i}{\partial x_k} + \frac{\partial v_k}{\partial x_i} \right) \tag{7.14}$$

and

$$e_{rr} = \frac{\partial v_r}{\partial x_r}, \tag{7.15}$$

tensor τ_{ik} becomes

$$\tau_{ik} = -P_0 \, \delta_{ik} + 2 \, \eta \, e_{ik} + \lambda \, \delta_{ik} \, e_{rr} . \tag{7.16}$$

Tensor e_{ik} is known as the *rate of strain tensor*, and Eq. (7.16) reflects the fact that the tension in a fluid is proportional to the deformation rate of the fluid. This result was initially obtained by Newton in the *Principia*. We should also note that τ_{ik} is an *isotropic tensor*, since the fluid was assumed to be isotropic, so that the following relations are valid

$$\begin{cases} \tau_{11} = -P_0 + 2 \, \eta \, e_{11} + \lambda \, e_{rr} \\ \tau_{22} = -P_0 + 2 \, \eta \, e_{22} + \lambda \, e_{rr} \\ \tau_{33} = -P_0 + 2 \, \eta \, e_{33} + \lambda \, e_{rr} , \end{cases} \tag{7.17}$$

or yet

$$\tau_{ii} = \tau_{11} + \tau_{22} + \tau_{33} , \tag{7.18}$$

so that

$$-\frac{1}{3} \tau_{ii} = -\frac{1}{3} \left[-3 \, P_0 + 2 \, \eta \, e_{rr} + 3 \, \lambda \, e_{rr} \right] = P_0 - \frac{2}{3} \, \eta \, e_{rr} - \lambda \, e_{rr} = P_0 - \left(\lambda + \frac{2}{3} \, \eta \right) e_{rr}, \tag{7.19}$$

where we have used a relation similar to (7.18) for the tensor e_{rr}. Let us define the average normal pressure P as

$$P = -\frac{1}{3} \tau_{ii} . \tag{7.20}$$

From (7.19) we have

$$P = P_0 - \left(\lambda + \frac{2}{3} \, \eta \right) e_{rr} . \tag{7.21}$$

Recalling that $e_{rr} = \nabla \cdot \mathbf{v}$ (see Eq. 7.15),

$$P = P_0 - \left(\lambda + \frac{2}{3} \eta \right) \nabla \cdot \mathbf{v} . \tag{7.22}$$

Again we see that, if $\mathbf{v} = 0$ or $\nabla \cdot \mathbf{v} = 0$, $P = P_0$, which is the static value of the tension. Otherwise, in a viscous fluid we have $P \neq P_0$, that is, the pressure defined by (7.20) is no longer a true thermodynamic variable, and depends on the rate of strain tensor. In order to prevent that, we must have

$$\begin{cases} \lambda + \dfrac{2}{3} \eta = 0 \\ \lambda = -\dfrac{2}{3} \eta , \end{cases} \tag{7.23}$$

so that $P = P_0$. Considering (7.23) and (7.9), the viscosity tensor σ_{ik} becomes

$$\sigma_{ik} = \eta \left(\frac{\partial v_i}{\partial x_k} + \frac{\partial v_k}{\partial x_i} - \frac{2}{3} \delta_{ik} \frac{\partial v_r}{\partial x_r} \right) \tag{7.24}$$

and the equation of motion (7.10) can be written as

$$\frac{\partial (\rho \, v_i)}{\partial t} = -\frac{\partial \Pi_{ik}}{\partial x_k} + \frac{\partial}{\partial x_k} \left[\eta \left(\frac{\partial v_i}{\partial x_k} + \frac{\partial v_k}{\partial x_i} - \frac{2}{3} \delta_{ik} \frac{\partial v_r}{\partial x_r} \right) \right]$$

$$\frac{\partial (\rho \, v_i)}{\partial t} = -\frac{\partial}{\partial x_k} \left[P \, \delta_{ik} + \rho \, v_i \, v_k - \eta \left(\frac{\partial v_i}{\partial x_k} + \frac{\partial v_k}{\partial x_i} - \frac{2}{3} \delta_{ik} \frac{\partial v_r}{\partial x_r} \right) \right] . \tag{7.25}$$

Equation (7.25) applies to the motion of a viscous fluid without external forces, such as the gravitational force, etc. Introducing an external force $\mathbf{F}$ (dyn/cm^3) characteristic of these processes, the equation becomes

$$\frac{\partial (\rho \, v_i)}{\partial t} = -\frac{\partial \Pi_{ik}}{\partial x_k} + \frac{\partial}{\partial x_k} \left[\eta \left(\frac{\partial v_i}{\partial x_k} + \frac{\partial v_k}{\partial x_i} - \frac{2}{3} \delta_{ik} \frac{\partial v_r}{\partial x_r} \right) \right] + F_i . \tag{7.26}$$

Recalling (7.2) we can write

$$\frac{\partial (\rho \, v_i)}{\partial t} = -\frac{\partial \Pi_{ik}}{\partial x_k} + \mathscr{F}_i + F_i \tag{7.27}$$

or

$$\frac{\partial (\rho \, v_i)}{\partial t} = -\frac{\partial}{\partial x_k} (\Pi_{ik} - \sigma_{ik}) + F_i . \tag{7.28}$$

It is usually possible to neglect the variation with position of the viscosity coefficients. In this case, η can be considered as a constant.

7.3   The Navier-Stokes Equation: Incompressible Fluids

Let us consider the particular case of a *incompressible fluid* under action of external forces. In this case, using Eqs. (1.9) and (5.23), we have

$$\begin{cases} \nabla \cdot \mathbf{v} = 0 \\[2mm] \dfrac{\partial v_r}{\partial x_r} = 0 , \end{cases} \tag{7.29}$$

Eq. (7.24) is reduced to

$$\sigma_{ik} = \eta \left(\frac{\partial v_i}{\partial x_k} + \frac{\partial v_k}{\partial x_i} \right) \tag{7.30}$$

and Eq. (7.26) becomes

$$\rho \frac{\partial v_i}{\partial t} = -\frac{\partial \Pi_{ik}}{\partial x_k} + \frac{\partial}{\partial x_k} \left[\eta \left(\frac{\partial v_i}{\partial x_k} + \frac{\partial v_k}{\partial x_i} \right) \right] + F_i$$

$$\rho \frac{\partial v_i}{\partial t} = -\frac{\partial \Pi_{ik}}{\partial x_k} + \eta \frac{\partial^2 v_i}{\partial x_k \, \partial x_k} + F_i , \tag{7.31}$$

where we have used the relation $\partial(\partial v_k/\partial x_i)/\partial x_k = \partial(\partial v_k/\partial x_k)/\partial x_i = 0$. Using (5.40) with $A_i = v_i$, we have the relations

$$\frac{D v_i}{Dt} = \frac{\partial v_i}{\partial t} + v_k \frac{\partial v_i}{\partial x_k} \tag{7.32}$$

$$\rho \frac{\partial v_i}{\partial t} = \rho \frac{D v_i}{Dt} - \rho \, v_k \frac{\partial v_i}{\partial x_k} . \tag{7.33}$$

Substituting (7.33) in (7.31) and using Eq. (6.10),

$$\rho \frac{D v_i}{Dt} = \rho \, v_k \frac{\partial v_i}{\partial x_k} - \frac{\partial(P \, \delta_{ik})}{\partial x_k} - \frac{\partial(\rho \, v_i \, v_k)}{\partial x_k} + \eta \frac{\partial^2 v_i}{\partial x_k \, \partial x_k} + F_i . \tag{7.34}$$

Using Eq. (6.8),

$$\rho \, v_k \frac{\partial v_i}{\partial x_k} - \frac{\partial(\rho \, v_i \, v_k)}{\partial x_k} = v_i \frac{\partial \rho}{\partial t} = 0 . \tag{7.35}$$

Therefore, (7.34) becomes

$$\rho \frac{Dv_i}{Dt} = -\frac{\partial (P\,\delta_{ik})}{\partial x_k} + \eta \frac{\partial^2 v_i}{\partial x_k\,\partial x_k} + F_i \,. \tag{7.36}$$

Coefficient η is the *dynamic viscosity coefficient*. We can also use the *kinematic viscosity coefficient* v,

$$v = \frac{\eta}{\rho}\,, \tag{7.37}$$

with units: $[v] = (\mathrm{dyn\,s\,cm^{-2}})\ (\mathrm{cm^3/g}) = \mathrm{dyn\,s\,cm\,g^{-1}} = \mathrm{cm^2/s}$. Equation (7.36) becomes

$$\begin{cases} \rho \dfrac{Dv_i}{Dt} = -\dfrac{\partial (P\,\delta_{ik})}{\partial x_k} + \rho v \dfrac{\partial^2 v_i}{\partial x_k\,\partial x_k} + F_i \\[4mm] \dfrac{Dv_i}{Dt} = -\dfrac{1}{\rho}\dfrac{\partial (P\,\delta_{ik})}{\partial x_k} + v \dfrac{\partial^2 v_i}{\partial x_k\,\partial x_k} + \dfrac{1}{\rho} F_i \,. \end{cases} \tag{7.38}$$

Writing Eqs. (7.36) and (7.38) in vector notation, we have

$$\begin{cases} \rho \dfrac{D\mathbf{v}}{Dt} = -\nabla P + \eta\,\nabla^2\mathbf{v} + \mathbf{F} \\[3mm] \rho \dfrac{D\mathbf{v}}{Dt} = -\nabla P + \rho v\,\nabla^2\mathbf{v} + \mathbf{F} \\[3mm] \dfrac{D\mathbf{v}}{Dt} = -\dfrac{1}{\rho}\,\nabla P + v\,\nabla^2\mathbf{v} + \dfrac{1}{\rho}\,\mathbf{F} \,. \end{cases} \tag{7.39}$$

Equations (7.31), (7.36), (7.38) and (7.39) are alternative forms of the *Navier-Stokes equation* in the case of incompressible fluids. This equation represents the application of Newton's second law to a viscous fluid.

The introduction of the viscosity tensor σ_{ik} in the Euler equation (see for example Eq. 7.4) also leads to a modification in the energy equation, since the energy lost has an additional component proportional to $v_i\,\partial\sigma_{ik}/\partial x_k$, with units: $\mathrm{erg\,cm^{-3}\,s^{-1}}$ (see Exercise 7.5 for an incompressible fluid).

7.4 The Navier-Stokes Equation: Compressible Fluids

Let us go back to Eq. (7.26) and obtain the Navier-Stokes equation for a compressible fluid in the presence of external forces. Using (7.26), (7.33) and Eqs. (6.4) and (6.10) we get

$$\rho \frac{Dv_i}{Dt} = -\frac{\partial(P\,\delta_{ik})}{\partial x_k} + \eta\,\frac{\partial^2 v_i}{\partial x_k\,\partial x_k} + \eta\,\frac{\partial^2 v_k}{\partial x_k\,\partial x_i} - \frac{2}{3}\,\eta\,\frac{\partial^2 v_r}{\partial x_k\,\partial x_r} + F_i \qquad (7.40)$$

and this equation can be written as

$$\begin{cases} \rho \dfrac{Dv_i}{Dt} = -\dfrac{\partial(P\,\delta_{ik})}{\partial x_k} + \eta\,\dfrac{\partial^2 v_i}{\partial x_k\,\partial x_k} + \dfrac{\eta}{3}\,\dfrac{\partial^2 v_r}{\partial x_k\,\partial x_r} + F_i \\[2mm] \rho \dfrac{Dv_i}{Dt} = -\dfrac{\partial(P\,\delta_{ik})}{\partial x_k} + \rho\,v\,\dfrac{\partial^2 v_i}{\partial x_k\,\partial x_k} + \dfrac{\rho\,v}{3}\,\dfrac{\partial^2 v_r}{\partial x_k\,\partial x_r} + F_i \\[2mm] \dfrac{Dv_i}{Dt} = -\dfrac{1}{\rho}\,\dfrac{\partial(P\,\delta_{ik})}{\partial x_k} + v\,\dfrac{\partial^2 v_i}{\partial x_k\,\partial x_k} + \dfrac{v}{3}\,\dfrac{\partial^2 v_r}{\partial x_k\,\partial x_r} + \dfrac{1}{\rho}\,F_i \ . \end{cases} \qquad (7.41)$$

Using vector notation we have

$$\begin{cases} \rho\,\dfrac{D\mathbf{v}}{Dt} = -\nabla P + \eta\,\nabla^2\mathbf{v} + \dfrac{\eta}{3}\,\nabla(\nabla\cdot\mathbf{v}) + \mathbf{F} \\[2mm] \rho\,\dfrac{D\mathbf{v}}{Dt} = -\nabla P + \rho\,v\,\nabla^2\mathbf{v} + \dfrac{\rho\,v}{3}\,\nabla(\nabla\cdot\mathbf{v}) + \mathbf{F} \\[2mm] \dfrac{D\mathbf{v}}{Dt} = -\dfrac{1}{\rho}\,\nabla P + v\,\nabla^2\mathbf{v} + \dfrac{v}{3}\,\nabla(\nabla\cdot\mathbf{v}) + \dfrac{1}{\rho}\,\mathbf{F} \ . \end{cases} \qquad (7.42)$$

In analogy to (7.36), (7.38), and (7.39), Eqs. (7.41) and (7.42) are the *Navier-Stokes equation for compressible fluids*. It should be noted that, if we do not use hypothesis (7.23), the coefficient $\eta/3$ in (7.41) or (7.42) must be replaced by $\eta/3 + \xi$, where $\xi = \lambda + (2/3)\,\eta$.

The introduction of the viscosity also changes the energy equation, since a new mechanism of energy dissipation is taken into account. A detailed discussion of the energy equation for viscous fluids is beyond the scope of this book, and the interested reader may consult the bibliography at the end of the chapter. A simple application example is given in Exercise 7.5.

Exercises

7.1. Show that, in order to reduce Eqs. (7.9)–(7.24), we must have $\lambda = -(2/3)\,\eta$.

7.2. Equation (7.9) is sometimes written as

$$\sigma_{ik} = \eta\left(\frac{\partial v_i}{\partial x_k} + \frac{\partial v_k}{\partial x_i} - \frac{2}{3}\,\delta_{ik}\,\frac{\partial v_r}{\partial x_r}\right) + \xi\,\delta_{ik}\,\frac{\partial v_r}{\partial x_r} \ .$$

What are the relations between the coeffficients (η, ξ) and coefficients (η, λ) defined by (7.9)?

7.3. Write the Navier-Stokes equation for compressible fluids in a form equivalent to Eq. (7.31).

7.4. Consider a viscous fluid with dynamic viscosity coefficient η in a flow characterized by velocity v, density ρ and linear dimension L. Show that the only dimensionless number that can be formed including these four properties is given by

$$R = \frac{v\,L\,\rho}{\eta}\ .$$

This is *Reynolds number*, essentially a measure of the ratio between the inertial and viscous forces acting on the fluid. Low values of R are associated with laminar flows, in which any perturbations are rapidly damped, while large values of R are applied to turbulent fluids.

7.5. Consider the energy equation as given by (6.29). Show that for an incompressible fluid characterized by the viscosity tensor σ_{ik}, this equation can be written as

$$\frac{\partial[\rho(v^2/2 + e)]}{\partial t} + \frac{\partial(\epsilon_k - v_i\,\sigma_{ik})}{\partial x_k} + \sigma_{ik}\,\frac{\partial v_i}{\partial x_k} = v_i\,F_i\ .$$

Bibliography

Kundu, P.K.: Fluid Mechanics. Academic, New York (2011) (Comprehensive book, including the basic equations of fluid dynamics, some applications and a revision of vector and tensor calculus)

Landau, L., Lifchitz, E.: Mécanique des Fluides. MIR, Moscou (1971) (English edition: Fluid Mechanics. Butterworth-Heinemann, Boston (1995). Referred to in Chapter 1. Presents a detailed treatment of the motion of viscous fluids and the Navier-Stokes equation, using both vector and cartesian tensor notation, which are written explicitly in spherical and cylindrical coordinates)

Trefil, J.S.: Introduction to the Physics of Fluids and Solids. Dover, New York (2010) (Referred to in Chapter 1. Includes a good qualitative discussion of fluid viscosity and the Navier-Stokes equation, some application examples and exercises)

Chapter 8
Sound Waves

Abstract This chapter presents an analysis of the propagation of sound waves. The wave equation is obtained and it is shown that the wave propagates with the speed of sound. A discussion is given on the existence of a critical point in the solution of the hydrodynamic equations in stellar winds in the isothermal case. The general topology of the possible solutions is presented and some consequences on the stellar mass loss rate are discussed.

8.1 Propagation of Sound Waves

Let us consider the propagation of *sound waves*, that is, low-amplitude vibrational motions in an ideal compressible fluid that propagate with the speed of sound. In this case, the velocity of the fluid particles is small compared to the wave velocity, and the variations of the pressure P and density ρ are also small. Let us restrict ourselves to a linear analysis, that is, variables such as the pressure and density can be written as

$$\begin{cases} P = P_0 + P' \\ \rho = \rho_0 + \rho' , \end{cases} \tag{8.1}$$

where P_0 and ρ_0 are the equilibrium values and P' and ρ' are the perturbations, or variations caused by the wave. The following relations are valid:

$$\begin{cases} P' \ll P_0 \\ \rho' \ll \rho_0 . \end{cases} \tag{8.2}$$

The continuity equation is given by (see Eq. 1.6):

$$\frac{\partial \rho}{\partial t} + \nabla \cdot (\rho \mathbf{v}) = 0 . \tag{8.3}$$

W.J. Maciel, *Hydrodynamics and Stellar Winds: An Introduction*, Undergraduate Lecture Notes in Physics, DOI 10.1007/978-3-319-04328-9_8,
© Springer International Publishing Switzerland 2014

Neglecting second order terms, this can be written as

$$\frac{\partial \rho'}{\partial t} + \rho_0 \, \nabla \cdot \mathbf{v} = 0 \, . \tag{8.4}$$

The Euler equation in the absence of external forces is (Eq. 2.13):

$$\frac{\partial \mathbf{v}}{\partial t} + (\mathbf{v} \cdot \nabla) \, \mathbf{v} = -\frac{1}{\rho} \, \nabla P \, . \tag{8.5}$$

Linearizing this equation and recalling that the term $(\mathbf{v} \cdot \nabla) \, \mathbf{v}$ is negligible in view of the small velocity $\mathbf{v}$, we get

$$\frac{\partial \mathbf{v}}{\partial t} + \frac{1}{\rho_0} \, \nabla P' = 0 \, . \tag{8.6}$$

The propagation of a sound wave in a perfect fluid is an *adiabatic* process, so that

$$P' = \left(\frac{\partial P}{\partial \rho_o} \right)_S \rho' \tag{8.7}$$

where the subscript S means constant entropy. Substituting (8.7) in (8.4), we have

$$\frac{\partial P'}{\partial t} + \rho_0 \left(\frac{\partial P}{\partial \rho_0} \right)_S \nabla \cdot \mathbf{v} = 0 \, . \tag{8.8}$$

Equations (8.6) and (8.8) involving the variables P' and $\mathbf{v}$ completely describe the sound wave. Let us now introduce the velocity potential φ (see Exercise 8.1) such that

$$\mathbf{v} = \nabla \varphi \, . \tag{8.9}$$

Using (8.9) and (8.6), we find

$$\frac{\partial (\nabla \varphi)}{\partial t} + \frac{\nabla P'}{\rho_0} = 0 \, ,$$

so that

$$\nabla P' = -\rho_0 \, \frac{\partial (\nabla \varphi)}{\partial t} = -\rho_0 \, \nabla \left(\frac{\partial \varphi}{\partial t} \right) = \nabla \left[-\rho_0 \left(\frac{\partial \varphi}{\partial t} \right) \right] ,$$

that is,

$$P' = -\rho_0 \, \frac{\partial \varphi}{\partial t} \, . \tag{8.10}$$

Substituting (8.10) in (8.8), and dropping the subscript in ρ_0, we get

$$\frac{\partial^2 \varphi}{\partial t^2} - \left(\frac{\partial P}{\partial \rho}\right)_S \nabla^2 \varphi = 0 ,$$

that can be written as

$$\frac{\partial^2 \varphi}{\partial t^2} - c_s^2 \nabla^2 \varphi = 0 , \tag{8.11}$$

which is a *wave equation*. Naturally, the term c_s is propagation velocity of the sound wave, or *speed of sound*, given by

$$c_s^2 = \left(\frac{\partial P}{\partial \rho}\right)_S . \tag{8.12}$$

The introduction of perturbations in a moving fluid as given by Eqs. (8.1) is related to the important issue of the fluid *stability*. Several physical processes may lead to the growth of perturbations in fluids, which also generate instabilities. In this book, these processes are not taken into account, and the interested reader may consult some of the texts in the bibliography at the end of the chapter.

8.2 The Speed of Sound in the Isothermal Case

Let us consider an *isothermal* transformation in a perfect gas. Analogously to Eq. (8.12), the sound speed is:

$$c_s^2 = \left(\frac{\partial P}{\partial \rho}\right)_T , \tag{8.13}$$

where the subscript T means constant temperature, that is

$$T = \text{constant} . \tag{8.14}$$

Using the equation of state of a perfect gas (Eq. 3.12),

$$P = \frac{k \rho T}{\mu m_H} \tag{8.15}$$

under condition (8.14), and assuming a constant molecular weight μ, we get

$$c_s^2 = \frac{k T}{\mu m_H} \tag{8.16}$$

or

$$c_s^2 = \frac{P}{\rho} \,.$$

(8.17)

8.3   The Speed of Sound in the Adiabatic Case

Let us obtain an expression for c_s in an *adiabatic* process in a perfect gas, if the equation of state is given by (8.15). We may replace the energy equation by a relation such as

$$P = \text{constant} \times \rho^\gamma \,,$$

(8.18)

where $\gamma = c_P/c_V$ is the ratio between the specific heats c_P and c_V. Therefore,

$$\left(\frac{\partial P}{\partial \rho} \right)_S = \gamma \, \frac{P}{\rho} \,.$$

(8.19)

Substituting in (8.12),

$$c_s^2 = \gamma \, \frac{P}{\rho} \,.$$

(8.20)

Considering Eq. (8.15), we may write

$$c_s^2 = \frac{\gamma \, k \, T}{\mu \, m_H} \,.$$

(8.21)

Usually γ and μ depend weakly on the temperature, so that we can consider $c_s \propto \sqrt{T}$ as a good approximation.

The relation between the adiabatic and isothermal compressibility is

$$\left(\frac{\partial P}{\partial \rho} \right)_S = \gamma \left(\frac{\partial P}{\partial \rho} \right)_T = \frac{c_P}{c_V} \left(\frac{\partial P}{\partial \rho} \right)_T \,.$$

(8.22)

In isothermal processes, apart from the (slow) variation with the molecular weight μ, the sound speed is constant. In an ideal fluid, without viscosity, conductivity, etc., the propagation of sound waves is adiabatic. When there is some energy transfer between the fluid elements and the surroundings, for example via conduction or radiation, the temperature fluctuations of the sound wave will be damped, and the propagation is isothermal at a speed given by (8.16) or (8.17).

8.4 The Critical Point in Stellar Winds

We can use Eq. (8.20) for the adiabatic speed of sound in order to examine an important aspect of the stellar wind theory, namely is the presence of a *critical point* in the hydrodynamic solutions. The critical point usually coincides or is located near the *sonic point*, where the flow velocity is equal to the sound speed in the gas. Let us consider a spherically symmetric stellar envelope and write the momentum conservation equation 2.73 in the form

$$v \frac{dv}{dr} = -\frac{1}{\rho} \frac{dP}{dr} - g_{ef} ,$$

(8.23)

where g_{ef} is the *effective gravity*, which takes into account the stellar gravity and any additional components that may act on the fluid. The mass conservation equation (1.19) can be written as

$$r^2 \rho v = \text{constant} .$$

(8.24)

Let us assume an adiabatic process, so that the gas pressure is related to the gas density by (8.18), where γ is the effective ratio of the specific heats, considered as a constant in a first approximation. Taking the derivative of (8.24) relative to r and simplifying, we have

$$\frac{2}{r} + \frac{1}{v} \frac{dv}{dr} + \frac{1}{\rho} \frac{d\rho}{dr} = 0$$

(8.25)

or

$$-\frac{1}{\rho} \frac{d\rho}{dr} = \frac{1}{v} \frac{dv}{dr} + \frac{2}{r} .$$

(8.26)

Differentiating (8.18),

$$\frac{dP}{dr} = \gamma \frac{P}{\rho} \frac{d\rho}{dr} = c_s^2 \frac{d\rho}{dr} ,$$

(8.27)

where we have introduced the adiabatic speed of sound from (8.20). Substituting (8.26) and (8.27) in (8.23), we have

$$v \frac{dv}{dr} = -\frac{1}{\rho} c_s^2 \frac{d\rho}{dr} - g_{ef} = c_s^2 \left(\frac{1}{v} \frac{dv}{dr} + \frac{2}{r} \right) - g_{ef}$$

$$\left(v - \frac{c_s^2}{v} \right) \frac{dv}{dr} = \frac{2 c_s^2}{r} - g_{ef} .$$

(8.28)

Multiplying both members of (8.28) by r we have

$$\left(v - \frac{c_s^2}{v}\right) r \frac{dv}{dr} = 2c_s^2 - g_{ef}\, r = \Delta \tag{8.29}$$

where we have introduced the function

$$\Delta = 2c_s^2 - g_{ef}\, r . \tag{8.30}$$

Isolating $1/v$ in (8.29), we obtain the relations:

$$\begin{cases} \dfrac{r}{v}\dfrac{dv}{dr} = \dfrac{\Delta}{v^2 - c_s^2} \\[2mm] \dfrac{d\ln v}{d\ln r} = \dfrac{\Delta}{v^2 - c_s^2} \end{cases} . \tag{8.31}$$

Equations (8.31) show a general characteristic of most stellar winds. At the sonic point, where $v = c_s$, there is a singularity in the momentum conservation equation, so that for realistic mass flows we must have $\Delta = 0$ at this point to keep the velocity gradient as a finite quantity.

8.4.1   *Example: Radiation Pressure on Molecules*

The outer atmospheres of red giant stars, in particular the Mira variables, are cool enough to allow the condensation of simple molecules, such as CO, H_2O and OH. The action of the stellar radiation pressure on molecular lines and bands may be an efficient driving mechanism for the mass loss observed in these stars. As an example, Fig. 8.1 shows a result by Maciel (1976) in the case of a red giant star wind driven by the action of the stellar radiation pressure on molecular lines. The sonic point is located near the star, where $r \simeq 1.083\, R$, with $\Delta = 0$ and a finite velocity gradient as the flow passes from the subsonic to the supersonic regime. Some values of Δ (km/s)^2 are shown at the right vertical scale of the figure. The mass loss rate calculated by this model is $dM/dt \simeq 1.3 \times 10^{-7}\, M_\odot/\text{year}$, suggesting that this mechanism may play an important role the mass loss process in cool giant stars.

8.5   Structure of Isothermal Winds

Some basic concepts of the stellar wind theory can be illustrated on the basis of very simple examples, as we have seen in Sect. 8.4. In the present section, we will show a somewhat more general treatment of the flow equations at the acceleration region and display the main solutions. Let us consider a simple case, in which the wind

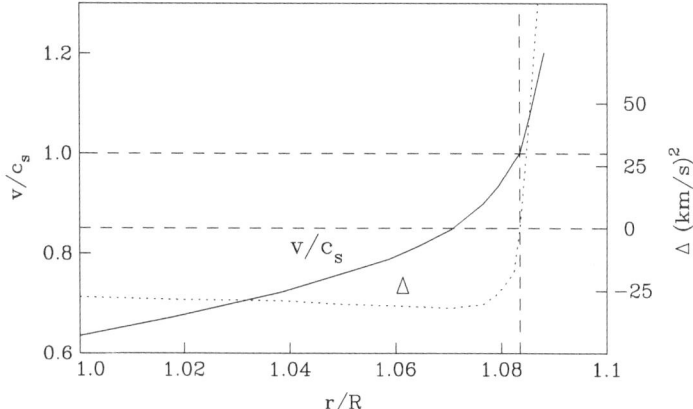

Fig. 8.1 A solution for the velocity law in red giant stars. The sonic point is indicated by the *vertical dashed line* (Maciel 1976)

may be considered as *isothermal, spherically symmetric,* and *time independent,* and the outwards force is essentially due to the gas pressure. Despite the simplicity of these hypotheses, several characteristics of the present analysis may also be applied to more detailed and realistic models.

The continuity equation can be written as

$$\frac{dM}{dt} = 4\pi r^2 \rho v \tag{8.32}$$

(see Eq. 1.29), and the momentum equation is

$$v\frac{dv}{dr} + \frac{1}{\rho}\frac{dP}{dr} + \frac{GM_*}{r^2} = 0 \tag{8.33}$$

(see Eq. 2.70). In this equation, the first term gives the gas acceleration (cm/s²), the second term is the force per unit mass due to the gas pressure, and the third term is the acceleration component due to the gravity of a star with mass M_*. The energy equation has the simplified form

$$T = \text{constant}, \tag{8.34}$$

which naturally means that some mechanism keeps a constant temperature throughout the stellar envelope. We can also use the equation of state (8.15), that is

$$P = \frac{k\rho T}{\mu m_H}, \tag{8.35}$$

where the mean molecular weight μ is assumed to be constant. Differentiating (8.35), we obtain the pressure gradient

$$\frac{1}{\rho}\frac{dP}{dr} = \frac{kT}{\mu m_H}\frac{1}{\rho}\frac{d\rho}{dr} , \qquad (8.36)$$

since the temperature is kept constant. Differentiating now (8.32), we obtain the density gradient

$$\frac{1}{\rho}\frac{d\rho}{dr} = -\frac{1}{v}\frac{dv}{dr} - \frac{2}{r} \qquad (8.37)$$

(see Eq. 8.26). Substituting (8.37) and (8.36) in the momentum equation (8.33), we get

$$v\frac{dv}{dr} + \frac{kT}{\mu m_H}\left(-\frac{1}{v}\frac{dv}{dr} - \frac{2}{r}\right) + \frac{G M_*}{r^2} = 0 . \qquad (8.38)$$

Using again the sound speed in the isothermal case given by (8.16), that is,

$$c_s = \left(\frac{kT}{\mu m_H}\right)^{1/2} , \qquad (8.39)$$

Eq. (8.38) can be written as

$$\frac{1}{v}\frac{dv}{dr} = \frac{\dfrac{2 c_s^2}{r} - \dfrac{G M_*}{r^2}}{v^2 - c_s^2} \qquad (8.40)$$

or

$$\frac{r}{v}\frac{dv}{dr} = \frac{d\ln v}{d\ln r} = \frac{\Delta}{v^2 - c_s^2} , \qquad (8.41)$$

where we have defined

$$\Delta = 2 c_s^2 - \frac{G M_*}{r} . \qquad (8.42)$$

We should note the similarity between Eqs. (8.41) and (8.31), and between the definitions of the Δ function in (8.30) and (8.42). The equation to be solved is then (8.40) or (8.41). At the base of the envelope, where $r = r_0$, which is usually considered as near the stellar photosphere, we must have $v(r_0) = v_0$. The numerator of the second member of (8.40) (Δ) is equal to zero at the position given by $r_c = G M_*/2 c_s^2$, which defines the *critical radius* r_c, or the *critical point*. This point must be located beyond the base of the envelope, so that we must have

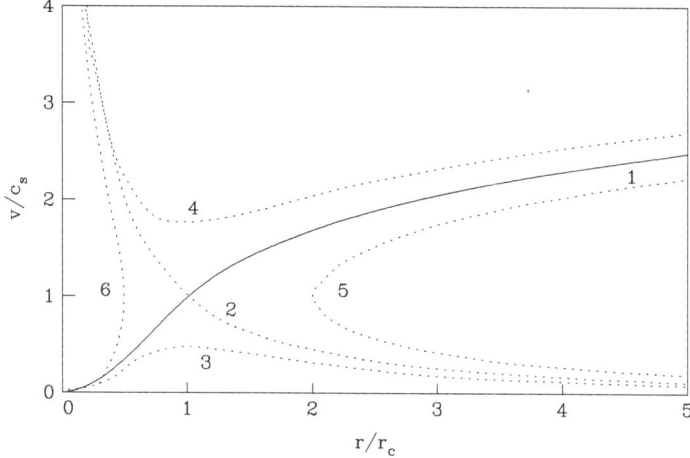

Fig. 8.2 Possible solutions of equation (8.40) for the velocity law in isothermal winds. *Curve 1* is the only one that goes through the critical point and reaches supersonic velocities

$$\frac{G M_*}{2 c_s^2} > r_0 ,\qquad(8.43)$$

that is,

$$\frac{G M_*}{2 r_0} > c_s^2 .\qquad(8.44)$$

At the critical point, the velocity gradient must be zero, unless $v(r_c) = c_s$. Analogously, if $v = c_s$, which defines the *sonic point*, the velocity gradient is infinity, unless this occurs at $r = r_c$. Therefore, *the only solution that can keep a positive velocity gradient throughout the envelope must pass through the critical point*, and is called the *critical solution*. In this case, we have the conditions

$$\begin{cases} r = r_c = \frac{G M_*}{2 c_s^2} \\ v(r) = v(r_c) = c_s . \end{cases}\qquad(8.45)$$

Near the base of the envelope, close to the stellar surface, the critical solution is subsonic, reaching supersonic velocities farther away from the star. In the isothermal case, the critical point and the sonic point coincide, which is not necessarily true in the general case, where the envelope has a different temperature profile, and other physical processes are in action.

Figure 8.2 shows the topology of the solutions of (8.40) or (8.41) for different values of the initial velocity v_0. Curve 1 is the critical solution, passing through the critical point from subsonic velocities up to supersonic velocities, and finally reaching the terminal velocity (not shown in the figure). Curve 2 also passes through the critical point, but the flow is initiated supersonically. Curve 3 is similar to the critical solution, but the initial velocity in not sufficient to keep the flow beyond the sonic point. On the other hand, curve 4 is entirely supersonic. Curves 5 and 6 have no physical meaning, since the solutions are not univocal. We see that, for $r < r_c$, $\Delta < 0$ and the velocity gradient is positive if $v < c_s$ (curves 1 and 3) and negative if $v > c_s$ (curves 2 and 4). For $r > r_c$, we have $\Delta > 0$ and the velocity gradient is positive if $v > c_s$ (curves 1 and 4) and negative if $v < c_s$ (curves 2 and 3).

The value of the velocity gradient of the critical solution at the sonic point can be obtained directly from (8.40) or (8.41) (see Exercise 8.5). The result is

$$\left(\frac{dv}{dr}\right)_{r_c} = \frac{2\,c_s^3}{G\,M_*}\,, \qquad (8.46)$$

which can be applied to curve 1 of Fig. 8.2, with a negative value of the same magnitude for curve 2.

The critical solution 1 is the only solution that passes through the critical point starting with subsonic velocities and reaching the supersonic phase. Therefore, the mass loss rate can be determined from this solution. Assuming the density at the base of the envelope to be ρ_0, we must have

$$\frac{dM}{dt} = 4\,\pi\,r_0^2\,\rho_0\,v_0\,, \qquad (8.47)$$

where v_0 is the initial velocity of the critical solution. Therefore, in the isothermal case, the location of the critical point and the mass loss rate depend essentially on the conditions at the subsonic region, and are independent of the terminal velocity of the gas.

For a star with mass M_*, radius R_* and molecular weight μ, knowing the parameters at the base of the envelope, r_0 and ρ_0, it is possible to obtain analytical expressions for the velocity law, the density distribution, and the mass loss rate as functions of temperature T, applying the transonic solution of equation (8.38). For example, adopting $M_* = 1\,M_\odot$, $R_* = 1\,R_\odot$, $\mu = 0.60$ and $\rho_0 = 1.0 \times 10^{-14}$ g/cm^3, Fig. 8.3 shows the mass loss rate dM/dt ($M_\odot$/year) as a function of the temperature for three different values of the radius at the base of the envelope, $r_0 = R_\odot$, $1.5\,R_\odot$ and $2.0\,R_\odot$. The dashed line represents the average mass loss rate of the Sun, $dM/dt \simeq 10^{-14}\,M_\odot$/year. We see that a coronal temperature $T \simeq 1.0 \times 10^6$ K produces results similar to the observations for $r_0 \simeq R_\odot$.

These results assume that the driving force of the wind is essentially due to the gas pressure. If additional components are in action, such as the radiative acceleration caused by ions in hot stars or molecules and grains in cool stars, the Euler equation must have additional terms involving the parameter Γ_r (see Eq. 2.77). In the subsonic region, this parameter has typical values in the range $0 < \Gamma_r < 1$, increasing the mass loss rate. In the supersonic region, it may reach higher values, which essentially affect the terminal velocities, but the mass loss rate is unchanged, since for isothermal winds it is fixed at the subsonic region.

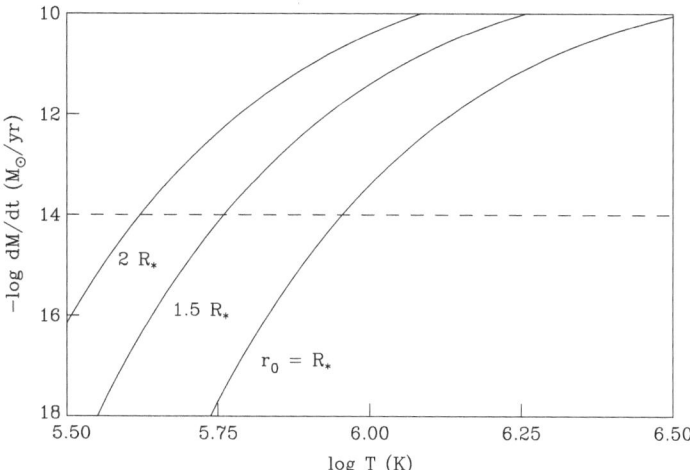

Fig. 8.3 Mass loss rates for solar-type stars as a function of temperature for isothermal winds. The *curves* refer to three values of the radius at the base of the envelope

Exercises

8.1. Show that condition (8.9) can be applied to the velocity field in an non-rotational flow. In this case, we have a potential flow, and the introduction of a velocity potential allows the use of scalar quantities instead of vector quantities.

8.2. In the example of Sect. 8.4, the average temperature in the envelope is $T \simeq$ 1,000 K and the mean molecular weight us $\mu \simeq 1.2$. (a) What is the gas velocity at the sonic point? Consider $\gamma \simeq 5/3$. (b) Knowing that the terminal velocity is $v_t \simeq 10$ km/s, what fraction of this velocity is reached at the sonic point?

8.3. Prove relations (8.40) and (8.41).

8.4. Show that in an isothermal wind under action of the pressure gradient the gas velocity at the critical point is given by

$$v(r_c) = 1/2 \ v_e(r_c)$$

where $v_e(r_c)$ is the escape velocity of the gas at the critical point.

8.5. Prove Eq. (8.46). Hint: Use de l'Hôpital rule to both numerator and denominator of Eq. (8.40).

Bibliography

Choudhuri, A.R.: The Physics of Fluids and Plasmas. Cambridge University Press, Cambridge (1998) (Referred to in Chapter 1. Presents a discussion of the structure of isothermal winds and the topology of the solutions of the wind equations)

Lamers, H.J.G.L.M., Cassinelli, J.P.: Introduction to Stellar Winds. Cambridge University Press, Cambridge (1999) (Referred to in Chapter 1. Excellent discussion of the stellar winds, in particular isothermal winds. Section 8.5 and Fig. 8.2 are based on this reference)

Landau, L., Lifchitz, E.: Mécanique des Fluides. MIR, Moscou (1971) (English edition: Fluid Mechanics. Butterworth-Heinemann, Boston (1995). Referred to in Chapter 1. Includes a good discussion of the propagation of sound waves in perfect and viscous fluids and of flow stability)

Maciel, W.J.: Mass loss from Mira variables by the action of radiation pressure on molecules. Astron. Astrophys. **48**, 27 (1976) (A study of mass loss in red giant stars driven by the stellar radiation pressure on molecular lines. Figure 8.1 is based on this reference. See also Jørgensen, U.G., Johnson, H.R., Astron. Astrophys. **265**, 168 (1992) and Olofsson, H., Phys. Scr. **T133** (2008))

Meyer-Vernet, N.: Basics of the Solar Wind. Cambridge University Press, Cambridge (2012) (Reprint of the 2007 edition of a modern introduction to the solar wind, with a discussion on the sun as a star, and the origin of the solar wind)

Parker, E.: Dynamics of the interplanetary gas and magnetic fields. Astrophys. J. **128**, 664 (1958) (The original solution of the isothermal wind equations applied to the solar wind is due to Eugene Parker. See also Brandt, J.C.: Introduction to the Solar Wind. Freeman, San Francisco (1970))

Chapter 9
Shock Waves

Abstract This chapter discusses shock waves, especially those generated by the interaction of stars and the interstellar medium through stellar winds. The Rankine-Hugoniot conditions are described, and some simple examples are presented for adiabatic and isothermal shocks.

9.1 Introduction

In a compressible fluid, the propagation rate of a perturbation is determined by the sound speed in the fluid. If the flow is sufficiently rapid, so that its velocity is larger than the sound speed, the fluid is not able to adjust itself by the propagation of sound waves, and its properties may change considerably over a short distance. In other words, if the pressure gradient determined by the wave is very large, that is, the physical conditions of the gas vary over a distance scale of the order of the mean free path of the gas molecules, microscopic processes will take place, originating viscous forces that act to decrease the pressure gradient. If the mean free path of the fluid particles is short compared to the typical dimensions of the system, we can imagine that the pressure variations in this region are discontinuous, that is, the fluid properties change in an essentially discontinuous way, creating a *shock wave*.

There are many astrophysical examples where shock waves occur. Pulsating stars, such as Cepheid variables and RR Lyrae stars generate waves in their inner photospheric layers, which propagate at supersonic velocities in the outer diffuse regions, producing shocks. Supernova explosions are extremely energetic phenomena, reaching energies of the order of 10^{50} erg, which are communicated to the surrounding gas by the expansion of the external layers of the stars at supersonic velocities, creating intense shock waves in the interstellar medium.

Supersonic velocities are also reached in the expansion of HII regions and planetary nebulae, which generate shock waves in the interaction region of the expanding layers with the interstellar medium (Fig. 9.1). According to the model that is generally accepted for the origin of planetary nebulae, the so-called interaction wind

W.J. Maciel, *Hydrodynamics and Stellar Winds: An Introduction*, Undergraduate
Lecture Notes in Physics, DOI 10.1007/978-3-319-04328-9_9,
© Springer International Publishing Switzerland 2014

Fig. 9.1 The planetary
nebula M27, a photoionized
expanding nebula, ejected by
a star with mass similar to
that of the Sun (Rodrigo
Campos, LNA)

model, shock waves are produced in the process of the formation of the nebulae.
These nebulae are formed during the late evolution stages of red giant stars with
masses close to the solar mass, or slightly larger. The expanding gas is ionized by
the photons emitted by the central star, which is a hot, collapsed object. In this case,
the hot and fast wind of the collapsed star generates shocks at it collides with the
cold, slow wind of the red giant progenitor star, thus creating the nebula, which can
be observed by intense emission lines. Filaments and other small scale structures
are also observed, frequently associated with internal shocks and X-ray emission.
Although this scheme seems to work for many nebulae, other processes are also
probably important, such as the presence of a magnetic field or the existence of a
binary companion star.

Generally speaking, stellar winds are also responsible for the propagation of
shock waves in the interstellar medium. This may occur in evolved stars, in main
sequence stars, in later evolution stages, or even in young objects, prior to the main
sequence, where jets associated to the shocks are frequently observed.

The detailed study of shocks and their astrophysical applications is beyond the
goals of this book, and the interested reader may look at some of the references
in the bibliography at the end of the chapter. We will consider here the main basic
characteristics of shocks in idealized situations, particularly in the case of stationary
shocks, as in steady flows associated with the interaction of stellar winds with the
interstellar medium, and neglecting dissipative processes involving the viscosity,
conduction, radiation, etc.

9.2   Shock Waves in the Interstellar Medium

The interaction of stellar winds with the interstellar medium is a good example of a
hydrodynamic process in a compressible fluid, with the production of shock waves.
This process leads to energy injection in an interstellar region with dimensions
considerably larger than the circumstellar dimensions associated with the mass
losing star. For example, a typical wind in a hot star has a terminal velocity of
$v \simeq 2,000$ km/s and a mass loss rate $dM/dt \sim 10^{-6} \, M_\odot$/year. In this case, the
mechanical energy injection rate in the interstellar gas is

Fig. 9.2 The Crab Nebula,
the remnant of a supernova
that exploded in 1054 AD
(Rodrigo Campos, LNA)

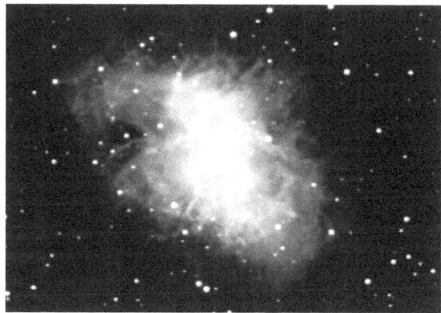

$$\frac{dE}{dt} \sim \frac{1}{2} \frac{dM}{dt} v_f^2 \sim 10^{36} \text{ erg/s} .$$

In this process, several layers with different properties are formed around the star:
near the star there is a wind moving at supersonic velocities relative to the interstellar
medium; in the next layer there is an interaction between the wind driven shock
waves and the gas at rest, and in the outer layers the gas remains unperturbed. In
the interaction region, the wind transfers energy and momentum to the gas, which is
heated to temperatures up to about 10^7 K, much higher than the temperature of the
unperturbed gas, which is typically of the order of 100 K. The gas cools basically
by emission of radiation, but part of the mechanical energy of the stellar wind is
transformed into kinetic energy of the gas particles. For a stellar wind associated
with a hot star of spectral type O, B, assuming spherical symmetry, neglecting the
radiative losses, and assuming that the interstellar gas is initially at rest, the radius r
of the affected region can be written approximately by

$$r \simeq 2.5 \times 10^{-19} \left[\frac{1}{\rho} \frac{dE}{dt} \right]^{1/5} t^{3/5} ,$$

where r is in parsec ($1 \text{ pc} = 3.1 \times 10^{18}$ cm), dE/dt is in erg/s, ρ in g/cm^3 and
the expansion timescale t is in seconds. Using typical values for hot stars, namely
$t \sim 10^5$ year and $dE/dt \sim 10^{36}$ erg/s, and adopting $\rho \sim 10^{-22}$ g/cm^3, we get
$r \simeq 3$ pc, that is, the affected region may reach a few parsecs beyond the star, where
the gas reaches velocities of the order of 20 km/s in a time scale of 10^5 year.

The estimates obtained for the stellar winds can be compared with the corresponding quantities in supernova explosions, when stars with masses larger than
about 10 $M_\odot$ explode, having consumed their nuclear fuel, transferring energies
of the order of 10^{49}–10^{51} erg to the interstellar medium. The explosions generate
intense shocks, heating and expanding the surrounding gas, in a process reaching
velocities of hundreds to thousands of km/s, sweeping about 10^3–10^4 $M_\odot$ of
interstellar material within several tens of parsecs, in timescales of hundreds to
hundreds of thousands years. The *supernova remnants* left over by the explosions
can be observed by their non-thermal emission. Figure 9.2 shows the Crab Nebula,
the remains of an explosion that was observed by Chinese astronomers in 1054 AD.

9.3 The Rankine-Hugoniot Conditions

Let us apply the basic hydrodynamic equations to a one-dimensional shock wave, neglecting any variations of the flow conditions during the time needed for the gas to cross the wave, that is, assuming the shock to be time independent. Let us consider a reference frame attached to the wave, so that the unperturbed gas has pressure P_0 and density ρ_0, moving with velocity u_0 relative to the shock (upstream gas), and the corresponding properties after the shock are P_1, ρ_1 and u_1 (downstream gas) (see Fig. 9.3). In this case, the mass conservation equation can be written considering that the mass flux of the gas approaching the shock (upstream gas) must be equal to the mass flux of the gas leaving the shock (downstream gas), as there is no creation or destruction of matter in the shock zone.

The mass flux in the pre-shock region is $\rho_0\, u_0$ ($\mathrm{g\,cm^{-2}\,s^{-1}}$), corresponding to the mass of gas that reaches the shock per unit area and per unit time, while in the post-shock region the equivalent quantity is $\rho_1\, u_1$. The mass conservation equation is written as

$$\rho_0\, u_0 = \rho_1\, u_1 \;. \tag{9.1}$$

In order to obtain the momentum conservation equation we should note that the momentum flux of the gas reaching the shock is $(\rho_0\, u_0)\, u_0 = \rho_0\, u_0^2$, with units $(\mathrm{g\,cm^{-3}\,cm\,s^{-1}})\,(\mathrm{cm\,s^{-1}}) = (\mathrm{g\,cm\,s^{-1}})\,(\mathrm{cm^{-2}\,s^{-1}})$. The corresponding quantity for the gas leaving the shock is $(\rho_0\, u_0)\, u_1$. The momentum change must be equal to the impulse of the resulting force, so that the momentum flux variation corresponds to the pressure difference in the two regions, that is

$$\rho_0\, u_0\, u_1 - \rho_0\, u_0^2 = P_0 - P_1 \;. \tag{9.2}$$

Using (9.1), Eq. (9.2) can be written as

$$P_0 + \rho_0\, u_0^2 = P_1 + \rho_1\, u_1^2 \;. \tag{9.3}$$

In order to obtain the energy equation, we should recall that there may be some energy transformation from one form to another at the shock front, but the total energy must be conserved, since there are no energy gains or losses when a gas mass reaches the shock front.

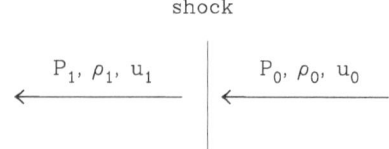

Fig. 9.3 Stationary shock. The gas approaches the shock from the *right* (upstream gas) and leaves the shock by the *left* (downstream gas)

Let us consider an "adiabatic" shock, that is, let us assume that the gas does not radiate during the shock. In fact, the gas entropy increases at the shock front, so that the gas is not adiabatic in the usual sense. The kinetic energy per unit mass of the unperturbed gas is $1/2\,u_0^2$, and its internal energy is

$$e_0 = \frac{1}{\gamma - 1}\frac{kT}{\mu\,m_H} = \frac{1}{\gamma - 1}\frac{P_0}{\rho_0}, \tag{9.4}$$

where we have considered a monatomic gas and a constant ratio of the specific heats at constant pressure and volume, $\gamma = c_P/c_V$. The energy per cm^2 per second entering the transition region is given by

$$E_0 = \rho_0\,u_0\left(\frac{1}{2}\,u_0^2 + \frac{1}{\gamma - 1}\frac{P_0}{\rho_0}\right), \tag{9.5}$$

where the first term gives the kinetic energy per unit mass and the second term is the corresponding internal energy. Analogously, for the region leaving the shock,

$$E_1 = \rho_1\,u_1\left(\frac{1}{2}\,u_1^2 + \frac{1}{\gamma - 1}\frac{P_1}{\rho_1}\right). \tag{9.6}$$

The difference between the energy before and after the shock must be equal to the work done by the gas, which results from the pressure difference. Therefore, the difference $E_1 - E_0$ (erg cm^{-2} s^{-1}) must be equal to the work done per unit area per unit time, that is,

$$\begin{cases} E_1 - E_0 = P_0\,u_0 - P_1\,u_1 \\ E_0 + P_0\,u_0 = E_1 + P_1\,u_1 . \end{cases} \tag{9.7}$$

Considering Eqs. (9.1) and (9.5)–(9.7), we get

$$\frac{1}{2}\,u_0^2 + \frac{\gamma}{\gamma - 1}\frac{P_0}{\rho_0} = \frac{1}{2}\,u_1^2 + \frac{\gamma}{\gamma - 1}\frac{P_1}{\rho_1}. \tag{9.8}$$

Adopting $\gamma = 5/3$, we obtain the well-known relation

$$\frac{1}{2}\,u_0^2 + \frac{5}{2}\frac{P_0}{\rho_0} = \frac{1}{2}\,u_1^2 + \frac{5}{2}\frac{P_1}{\rho_1}. \tag{9.9}$$

Equations (9.1), (9.3) and (9.8) or (9.9) express the physical conditions at the boundary region, and are the so-called *Rankine-Hugoniot conditions*, or *jump conditions*. We see that they express essentially the conservation of mass, momentum and energy in the discontinuity zone.

9.3.1   Example: Pressure in the Interstellar Clouds

The densities in the interstellar regions are very low, usually much lower than the best laboratory vacuum that can be produced. Also, there is a considerable variation in the densities of the different interstellar regions, as for example in the diffuse clouds, the molecular clouds or the photoionized nebulae (planetary nebulae and HII regions). Since the temperatures of these regions are usually in the range $10^4 > T(\mathrm{K}) > 10$, the corresponding pressures are also very low. For example, a typical diffuse interstellar cloud has density $n \simeq 10\,\mathrm{cm}^{-3}$ and temperature $T \simeq 100\,\mathrm{K}$. In this case, the cloud pressure is of the order of $P \simeq nkT \sim 10^{-13}\,\mathrm{dyn/cm}^2 \sim 10^{-19}\,\mathrm{atm} \sim 10^{-16}\,\mathrm{Torr}$. The pressure in a laboratory vacuum is of the order of $10^{-6}\,\mathrm{dyn/cm}^2 \sim 10^{-12}\,\mathrm{atm} \sim 10^{-9}\,\mathrm{Torr}$. In a dense interstellar cloud, such as a molecular cloud, the particle density may reach $n \sim 10^6\,\mathrm{cm}^{-3}$ at $T \simeq 30\,\mathrm{K}$, so that $P \sim 4 \times 10^{-9}\,\mathrm{dyn/cm}^2 \sim 4 \times 10^{-15}\,\mathrm{atm} \sim 4 \times 10^{-12}\,\mathrm{Torr}$. In the denser parts of HII regions or planetary nebulae we may have $n \sim 10^4\,\mathrm{cm}^{-3}$ at $T \sim 10^4\,\mathrm{K}$, so that $P \sim 10^{-8}\,\mathrm{dyn/cm}^2 \sim 10^{-14}\,\mathrm{atm} \sim 10^{-11}\,\mathrm{Torr}$.

9.4   The Mach Number

We can use Eqs. (9.1), (9.3) and (9.9) and write the Rankine-Hugoniot conditions directly in terms of the ratios between the pressure in regions 0 and 1, defining the *Mach Number* as

$$M_0 = \frac{u_0}{c_s}, \tag{9.10}$$

where the adiabatic sound speed c_s is given by

$$c_s^2 = \gamma \frac{P_0}{\rho_0} = \frac{5}{3} \frac{P_0}{\rho_0}, \tag{9.11}$$

adopting again $\gamma = 5/3$. We have then

$$M_0^2 = \frac{3}{5} \frac{u_0^2 \rho_0}{P_0}. \tag{9.12}$$

From (9.3),

$$\frac{P_1}{P_0} = 1 + \frac{\rho_0 u_0^2}{P_0} - \frac{\rho_1 u_1^2}{P_0}. \tag{9.13}$$

Using (9.1), (9.12) and (9.13), we get

$$\frac{P_1}{P_0} = 1 + \frac{5}{3} M_0^2 \left(1 - \frac{\rho_0}{\rho_1} \right). \tag{9.14}$$

Using now (9.1), (9.9) and (9.12), we can write

$$\frac{\rho_0 \, P_1}{\rho_1 \, P_0} - 1 = \frac{1}{3} \, M_0^2 \left[1 - \left(\frac{\rho_0}{\rho_1} \right)^2 \right]. \tag{9.15}$$

Equations (9.14) and (9.15) are alternative forms of the Rankine-Hugoniot equations. For any value of the γ ratio, it is easy to show that these equations become

$$\frac{P_1}{P_0} = 1 + \gamma \, M_0^2 \left(1 - \frac{\rho_0}{\rho_1} \right) \tag{9.16}$$

$$\frac{\rho_0 \, P_1}{\rho_1 \, P_0} - 1 = \frac{\gamma - 1}{2} \, M_0^2 \left[1 - \left(\frac{\rho_0}{\rho_1} \right)^2 \right]. \tag{9.17}$$

Equation (9.16) or (9.14) gives the pressure P_1 as a function of ρ_1, if P_0 and ρ_0 are known. It can be shown that the physically acceptable solution requires $P_1 > P_0$ and $\rho_1 > \rho_0$, that is, the shock compresses the gas.

Applying the Rankine-Hugoniot conditions (9.1), (9.3) and (9.9), we can derive some expressions for quantities such as the internal energy e_i, the kinetic energy, the total energy, the entropy and the Mach number in the shocked region. Considering an adiabatic flow in a perfect, monatomic gas, the entropy per unit mass is proportional to e_i/T. Since $T \propto \rho^{2/3}$ (see Eq. 4.25) and $e_i \propto P/\rho$, the function $P/\rho^{5/3}$ represents the entropy, which must increase in the shocked region. Since the total energy is conserved, the internal energy must increase through the transition zone at the expense of the decreasing kinetic energy. Consequently, the Mach number also decreases, that is $M > 1$ in the pre-shock region and $M < 1$ in the post-shock region. Therefore, the gas approaches the shock supersonically, and leaves the shock subsonically. The mathematical formalism including the variations of the physical conditions in the shocked zone can be found in the book by Dyson and Williams (1997, see also Exercise 9.4).

9.4.1 Example: The Mach Number in Interstellar Clouds

We can use the data of the previous example and Eq. (9.11) to estimate the sound speed in typical interstellar regions. For diffuse clouds we get $c_s \simeq 1.2$ km/s adopting $\gamma = 5/3$. For dense molecular clouds, assuming a mean molecular weight $\mu \simeq 2$, we get $c_s \simeq 0.5$ km/s. For photoionized nebulae (HII regions and planetary nebulae), with $\mu \simeq 1/2$, we have $c_s \simeq 17$ km/s. These velocities are generally much smaller than typical wind velocities, as we have seen. The winds of red giant stars expand towards the diffuse interstellar gas at velocities typically of $v \simeq 10$–20 km/s. In the case of hot, young stars, which are frequently associated with HII regions, or for the central stars of planetary nebulae, which are also associated with photoionized nebulae, the wind velocities are much higher, $v \simeq 1,000$–2,000 km/s.

Therefore, we have then $v \gg c_s$, so that shock waves are a natural consequence of the expansion process of the gas in the outer stellar atmospheres.

Assuming a stationary shock, the velocity of the gas approaching the shock is $u \simeq v$, so that the Mach number in the examples above is typically $M \simeq 10 - 100$. For example, in the case of a diffuse interstellar cloud having $n \simeq 10 \, \text{cm}^{-3}$, $T \simeq 100 \, \text{K}$ and $\mu \simeq 1$, we have $\rho_0 \simeq n_H \, m_H \simeq 1.7 \times 10^{-23} \, \text{g/cm}^3$, $P_0 \simeq n k T \simeq 1.4 \times 10^{-13} \, \text{dyn/cm}^2$, $c_s \simeq 1.2 \, \text{km/s}$ and $P_0 + \rho_0 \, u_0^2 \simeq 6.8 \times 10^{-11} \, \text{dyn/cm}^2$, that is, $M \simeq 17$ and $\rho_0 \, u_0^2 \gg P_0$.

9.5   Physical Conditions in the Shock Zone

Generally we would like to know the values of the physical properties such as the velocity, density, pressure and temperature in the post-shock region, u_1, ρ_1, P_1 and T_1 in terms of the corresponding values in the unperturbed region, u_0, ρ_0, P_0 and T_0. Let us rewrite Eq. (9.5) in the case of a shock where $\rho_0 \, u_0^2 \gg P_0$, as we have seen in Example 9.4.1. This condition corresponds to an intense shock, where $M_0 \gg 1$, as can be seen from Eq. (9.12). We have

$$P_1 + \rho_1 \, u_1^2 = P_0 + \rho_0 \, u_0^2 \simeq \rho_0 \, u_0^2 \, . \tag{9.18}$$

Using the conservation energy equation in the form (9.9) ($\gamma = 5/3$), we have

$$\frac{1}{2} u_1^2 + \frac{5}{2} \frac{P_1}{\rho_1} = \frac{1}{2} u_0^2 + \frac{5}{2} \frac{P_0}{\rho_0} \simeq \frac{1}{2} u_0^2 \, . \tag{9.19}$$

Replacing P_1 given by (9.18) in relation (9.19) and using the conservation of mass flux condition (9.1), we obtain the relation

$$\frac{1}{2} u_1^2 + \frac{5}{2} \left(\frac{\rho_0 \, \rho_1^2 \, u_1^2}{\rho_0^2 \, \rho_1} - u_1^2 \right) \simeq \frac{1}{2} \left(\frac{\rho_1}{\rho_0} \right)^2 u_1^2 \, . \tag{9.20}$$

Simplifying this expression and considering $u_1 \neq 0$, we obtain an equation for the density ratio ρ_1/ρ_0 given by

$$\left(\frac{\rho_1}{\rho_0} \right)^2 - 5 \left(\frac{\rho_1}{\rho_0} \right) + 4 = 0 \, . \tag{9.21}$$

The roots of this equation are $\rho_1/\rho_0 = 4$ and $\rho_1/\rho_0 = 1$. Therefore, for an intense shock, the density increases by a factor of 4 relative to the density of the unperturbed gas, that is,

$$\frac{\rho_1}{\rho_0} \simeq 4 \, . \tag{9.22}$$

This is the *compression ratio*. It does not depend on M_0, since an increase in the fluid kinetic energy before the shock corresponds to a larger translation energy after the shock, which limits compression. From (9.1) we see that

$$\frac{u_1}{u_0} \simeq \frac{1}{4},$$ (9.23)

that is, the gas leaves the shock with a speed four times slower. Since $E_c \propto u_1^2$, we see that, in fact, the gas kinetic energy decreases, compensating thus the increase in the internal energy. For intense shocks, we can use (9.1), (9.18), and (9.23) and write for the pressure P_1

$$P_1 \simeq \frac{3}{4} \, \rho_0 \, u_0^2 \, .$$ (9.24)

Since $\rho_0 \, u_0^2 \gg P_0$ and $P_1 \simeq (3/4) \, \rho_0 \, u_0^2$, we see that $P_1 \gg P_0$.

In order to obtain the gas temperature in the post-shock region, we can use the equation of state in the form $P_1 = k \, \rho_1 T_1 / \mu \, m_H$, replacing ρ_1 and P_1 by Eqs. (9.22) and (9.24), respectively. The result is

$$T_1 \simeq \frac{3 \, \mu \, m_H}{16 \, k} \, u_0^2 \, .$$ (9.25)

The relations (9.22)–(9.25) can be applied to intense shocks with $\gamma = 5/3$. For the general case the reader may consult the bibliography at the end of the chapter, especially Spitzer (1978), Landau and Lifchitz (1971) and Zeldovich and Raizer (2003).

9.5.1 Example: Diffuse Interstellar Clouds

Let us consider again the case of a diffuse interstellar cloud seen at the end of Example 9.4.1. With $T \simeq 100 \, K$, $\rho_0 \simeq 1.7 \times 10^{-23} \, g/cm^3$ and $u_0 \simeq 20 \, km/s$, we have from (9.22) $\rho_1 \simeq 4 \, \rho_0 \simeq 6.8 \times 10^{-23} \, g/cm^3$; from (9.23), $u_1 \simeq u_0/4 \simeq 5 \, km/s$; from (9.24) $P_1 \simeq 5.1 \times 10^{-11} \, dyn/cm^2$, and from (9.25), $T_1 \simeq 9.1 \times 10^3 \, K$, that is, $T_1 \gg T_0$, corresponding to the fact that the internal energy increases at the shock front.

9.6 Fixed Reference System

The results of the previous section were obtained for a reference frame centered in the shock wave, so that the shock remains stationary. We are frequently interested in a fixed reference frame, in which case both the shock and the gas are moving (Fig. 9.4). Let us consider now that u_0 is the velocity of the unperturbed gas

Fig. 9.4 Fixed reference
system. The pre-shock
velocity (*right*) and
post-shock velocity (*left*) are
indicated

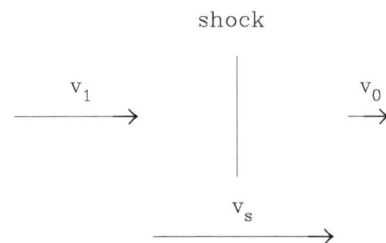

(*upstream gas*) relative to the shock; v_0 is the velocity of the unperturbed gas relative
to the fixed reference system; v_s is the velocity of the shock wave relative to the fixed
reference system; u_1 is the gas velocity in the post-shock region (*downstream gas*)
relative to the shock, and v_1 is the gas velocity in the post-shock region relative to
the fixed reference system.

We have then the following transformation relations:

$$\begin{cases} u_0 = v_0 - v_s \\ u_1 = v_1 - v_s \, . \end{cases} \tag{9.26}$$

Since we are interested in intense shocks, let us assume that $v_s \gg v_0$. In this case,
from (9.23) we have

$$\frac{v_1 - v_s}{v_0 - v_s} = \frac{1}{4} \simeq \frac{v_s - v_1}{v_s}$$

and

$$v_1 \simeq \frac{3}{4} v_s \, . \tag{9.27}$$

Therefore, for intense shocks the compressed gas moves in the same direction as
the shock wave, at 3/4 of its velocity, thus forming a compressed region behind the
shock. In the new reference system, Eq. (9.22) is conserved, that is, $\rho_1/\rho_0 \simeq 4$. The
pressure P_1 given by (9.24) is then

$$P_1 \simeq \frac{3}{4} \rho_0 v_s^2 \, , \tag{9.28}$$

and the temperature T_1 given by (9.25) becomes

$$T_1 \simeq \frac{3 \mu \, m_H}{16 \, k} v_s^2 \, . \tag{9.29}$$

Analogously, the internal energy $(3k T_1/2\mu m_H)$ and the kinetic energy $(v_1^2/2)$ in
the post-shock region are equal to $(9/32) \, v_s^2$.

9.6.1 Example: Winds in Hot Stars

Let us consider a stellar wind in a hot star that generates a shock wave with a velocity near the gas terminal velocity, $v_s \simeq 10^3$ km/s. Assuming that the wind propagates through an ionized hydrogen cloud where $\mu = 1/2$, $n_0 = 100$ cm^{-3} and $T_0 = 10^4$ K, we have in the pre-shock region the density $\rho_0 \simeq n_0\,\mu\,m_H \simeq 8.35 \times 10^{-23}$ g/cm^3 and a gas pressure of $P_0 \simeq k\,\rho_0\,T_0/\mu\,m_H \simeq 1.38 \times 10^{-10}$ dyn/cm^2. Assuming again $\gamma = 5/3$, the speed of sound in the gas is $c_s = (\gamma\,P_0/\rho_0)^{1/2} \simeq 16.6$ km/s. We see that $\rho_0\,u_0^2 \simeq \rho_0\,v_s^2 \gg P_o$, or $M_0 \simeq v_s/c_s \simeq 60 \gg 1$. The compressed gas density is $\rho_1 \simeq 4\,\rho_0 \simeq 3.34 \times 10^{-22}$ g/cm^3. From Eq. (9.28), the post-shock pressure is $P_1 \simeq (3/4)\,\rho_0\,v_s^2 \simeq 6.26 \times 10^{-7}$ dyn/cm^2, or $P_1 \gg P_0$. From (9.29), the post-shock temperature is $T_1 \simeq (3\,\mu\,m_H/16\,k)\,v_s^2 \simeq 1.13 \times 10^7$ K, that is, we have again $T_1 \gg T_0$.

9.7 Isothermal Shocks

Until now, we have assumed adiabatic shocks, where there is no radiation emission. In some cases, however, the compressed gas is able to radiate very efficiently, thus cooling rapidly. A relatively thin cooling layer is formed, followed by a more extended post-shock region where the temperature is essentially the same as in the pre-shock region, that is, $T_2 = T_0$ (see Fig. 9.5).

Assuming that the gas cools rapidly, the quantities P_0, ρ_0, u_0, and T_0 reach the values P_2, ρ_2, u_2 and $T_2 = T_0$ in the post-shock region. Since $c_s^2 \propto T$, the sound speed in the post-shock region is the same, that is,

$$c_{s2}^2 = c_{s0}^2 = \frac{P_0}{\rho_0} = \frac{P_2}{\rho_2}\,. \tag{9.30}$$

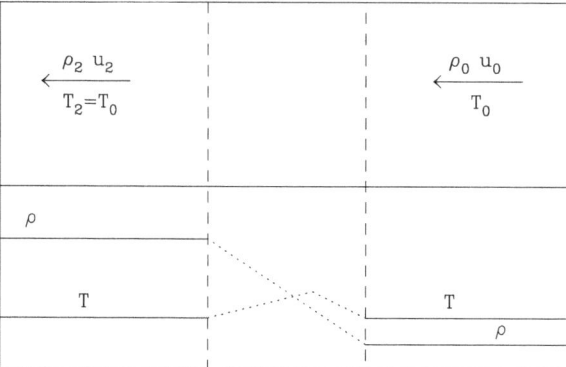

Fig. 9.5 Schematic representation of an isothermal shock. In the post-shock region the gas cools rapidly, emitting radiation and returning to its initial temperature

The mass flux and the momentum conservation equations can be written as

$$\rho_2\, u_2 = \rho_0\, u_0 \tag{9.31}$$

and

$$P_2 + \rho_2\, u_2^2 = P_0 + \rho_0\, u_0^2 , \tag{9.32}$$

respectively. Assuming again an intense shock, where $\rho_0\, u_0^2 \gg P_0$, from Eqs. (9.30) to (9.32) we have the equation

$$u_2^2 - u_0\, u_2 + c_s^2 = 0 , \tag{9.33}$$

where we have dropped the subscript of the speed of sound. The solution of (9.33) can be written as

$$u_2 = \frac{1}{2}\, u_0 \left[1 \pm \left(1 - \frac{4\, c_{s0}^2}{u_0^2} \right)^{1/2} \right] . \tag{9.34}$$

We have seen that $\rho_0\, u_0^2 \gg P_0$ corresponds to $M_0 \gg 1$ or $u_0 \gg c_s$, so that the term within parentheses in Eq. (9.45) can be expanded in series as

$$\left(1 - \frac{4\, c_s^2}{u_0^2} \right)^{1/2} \simeq 1 - \frac{2\, c_s^2}{u_0^2} .$$

Substituting this equation in (9.34), we obtain the two solutions of (9.33): the trivial solution $u_2 \simeq u_0$, corresponding to the case where there is no shock, and the nontrivial, adopted solution,

$$u_2 \simeq \frac{c_s^2}{u_0} . \tag{9.35}$$

Using this solution we can determine the compression ratio ρ_2/ρ_0 from Eq. (9.31),

$$\frac{\rho_2}{\rho_0} = \frac{u_0}{u_2} \simeq \frac{u_0^2}{c_s^2} = M_0^2 , \tag{9.36}$$

that is, the compression ratio of an isothermal shock is no longer limited to a factor of 4, as in the adiabatic case, but depends on the Mach number of the unperturbed gas. In order to keep a constant temperature, part of the internal energy is radiated, thus favouring compression.

We can use Eqs. (9.32), (9.35), and (9.36) to obtain the gas pressure in the post-shock region, with the result

$$P_2 \simeq \rho_0 u_0^2 \left(1 - \frac{c_s^2}{u_0^2}\right) \simeq \rho_0 u_0^2 \,. \tag{9.37}$$

For a fixed reference system, we have the relations $u_0 = v_0 - v_s$ and $u_2 = v_2 - v_s$ (see Eq. 9.26), with $v_s \gg v_0$. Using (9.36) we get

$$v_2 = v_s \left(1 - \frac{1}{M_0^2}\right) \simeq v_s \,, \tag{9.38}$$

since $M_0 \gg 1$, that is, the gas has approximately the same velocity and direction as the shock. Therefore, the compressed gas layer moves along with the shock front. The compression ratio is given by (9.36), and from (9.37) the post-shock pressure is

$$P_2 \simeq \rho_0 u_0^2 \simeq \rho_0 (v_0 - v_s)^2 \simeq \rho_0 v_s^2 \,. \tag{9.39}$$

In Fig. 9.5 we can see a schematic representation of the variations of the temperature T and density ρ for isothermal shocks. In the shocked region, the temperature increases up to a certain limiting value, decreasing thereafter to the original value, while the density increases up to the post-shock region.

9.7.1 Example: Isothermal Winds in Hot Stars

Let us consider again the previous Example 9.6.1 of shocks in the winds of hot stars, assuming now that the shock is isothermal, that is, the gas is allowed to emit radiation efficiently after the shock. Using data from the example and Eqs. (9.36), (9.38) and (9.39), we have the following results: the post-shock density is $\rho_2 \simeq M_0^2 \rho_0 \simeq 5.05 \times 10^{-19}$ g/cm^3, or $M_0 \simeq 78$; the gas velocity becomes $v_2 \simeq v_s \simeq 10^3$ km/s, and the post-shock pressure is $P_2 \simeq \rho_0 v_s^2 \simeq 8.35 \times 10^{-7}$ dyn/cm^2, about 4/3 of the corresponding value in the case of an adiabatic shock.

9.8 Hydromagnetic Shocks

Let us consider the propagation of a shock wave through a non-viscous medium subject to the action of a magnetic field $\mathbf{B}$. The Euler equation in the form (2.21) without external forces can be written as

$$\frac{\partial \mathbf{v}}{\partial t} + (\mathbf{v} \cdot \nabla) \mathbf{v} = -\frac{1}{\rho} \nabla P - \frac{1}{8\pi\rho} \nabla B^2 + \frac{1}{4\pi\rho} (\mathbf{B} \cdot \nabla) \mathbf{B} \,. \tag{9.40}$$

Assuming a uniform magnetic field, perturbations normal to the field (hydromagnetic waves, or Alfvén waves) propagate along $\mathbf{B}$ at the Alfvén velocity,

$$v_A = \left(\frac{B^2}{4\pi\rho}\right)^{1/2} . \tag{9.41}$$

In laboratory conditions, the velocity v_A is usually small compared to the sound speed. In some astrophysical situations, however, this does not happen, in particular in regions with low gas density. For example, in the chromospheres of cool giant stars, where $\rho \sim 10^{-13}\,\mathrm{g/cm^3}$, we get $v_A \simeq 90\,\mathrm{km/s}$ with a magnetic field $B \sim 10\,\mathrm{G}$, while $c_s \simeq 100\,\mathrm{km/s}$ for chromospheric temperatures $T \sim 10^4\,\mathrm{K}$. Hydromagnetic waves are observed in the solar wind, and probably participate in the process of momentum transfer to the expanding gas.

Considering again the unidimensional case, let us assume that the magnetic field lines are parallel to the shock front, taken as perpendicular to the x axis. In this case, the term $(\mathbf{B} \cdot \nabla)\mathbf{B}$ in Eq. (9.40) is equal to zero, and the equation of motion can be written as

$$\frac{\partial \mathbf{v}}{\partial t} + (\mathbf{v} \cdot \nabla)\,\mathbf{v} = -\frac{1}{\rho}\,\nabla P - \frac{1}{8\pi\rho}\,\nabla B^2 . \tag{9.42}$$

Integrating (9.42) in the shock zone, it is easy to see that the equivalent form of Eq. (9.3) is

$$P_0 + \rho_0\,u_0^2 + \frac{B_0^2}{8\pi} = P_1 + \rho_1\,u_1^2 + \frac{B_1^2}{8\pi} . \tag{9.43}$$

The magnetic flux remains constant in any trajectory through a conducting fluid, so that in the one-dimensional case we have

$$\frac{B_0}{\rho_0} = \frac{B_1}{\rho_1} . \tag{9.44}$$

9.8.1 Example: Adiabatic Interstellar Shock

We have seen in Sect. 9.5 that a non-radiating intense shock ("adiabatic" shock), the compression factor is $\rho_1/\rho_0 = 4$. In this case, the magnetic pressure varies from $P_{m0} = B_0^2/8\pi$ to $P_{m1} = B_1^2/8\pi$, that is,

$$P_{m1} = \frac{B_1^2}{8\pi} = \frac{B_0^2}{8\pi}\left(\frac{\rho_1}{\rho_0}\right)^2 = 16\,P_{m0} . \tag{9.45}$$

For example, in typical interstellar conditions, we have $B_0 \simeq 3 \times 10^{-6}\,\mathrm{G}$, $u_0 \simeq 10\,\mathrm{km/s}$ and $\rho_0 \simeq n\,m_H \simeq 2 \times 10^{-23}\,\mathrm{g/cm^3}$, with about $n \simeq 10$ particles per cubic centimeter. In this case, $P_{m0} \simeq 3.6 \times 10^{-13}$ dyn/cm^2 and $P_{m1} \simeq 5.7 \times 10^{-12}$ dyn/cm^2. From (9.24), we see that the gas pressure is of the order of $P_1 \simeq (3/4) \rho_0 u_0^2 \simeq 2 \times 10^{-11}$ dyn/cm^2, that is, $P_1 \gg P_{m1}$. Therefore, if the magnetic field in the unperturbed region is lower or of the order of $3\,\mu$G, the magnetic pressure in the post-shock region is negligible compared with the gas pressure in typical interstellar clouds.

9.8.2 Example: Isothermal Interstellar Shock

Let us consider now the effect of a magnetic field in an intense isothermal shock. We get from Eq. (9.43),

$$c_s^2 + u_0^2 + \frac{1}{2} v_{A0}^2 = \frac{\rho_1}{\rho_0} c_s^2 + \frac{u_0^2}{(\rho_1/\rho_0)} + \frac{1}{2} v_{A0}^2 \left(\frac{\rho_1}{\rho_0} \right)^2, \qquad (9.46)$$

where we have used Eqs. (9.30), (9.31), (9.41) and (9.44), adopting the indices "0" and "1" for the pre- and post-shock conditions, respectively. In principle, Eq. (9.46) must be solved in order to obtain the compression ratio ρ_1/ρ_0. Assuming the limiting case, where $\rho_1/\rho_0 \gg 1$ and $v_{A0} \gg c_s$, we have

$$u_0^2 \simeq \left(\frac{\rho_1}{\rho_0} \right)^2 \frac{v_{A0}^2}{2}, \qquad (9.47)$$

that is, the compression ratio becomes

$$\frac{\rho_1}{\rho_0} \simeq \sqrt{2} \frac{u_0}{v_{A0}}. \qquad (9.48)$$

Considering again the interstellar values of Example 9.8.1, $\rho_0 \simeq 2 \times 10^{-23}$ g/cm^3 and $B_0 \simeq 3\,\mu$G, we get from (9.41) $v_{A0} \simeq 1.9$ km/s. The speed of sound is close to this value, $c_s \simeq (kT/m_H)^{1/2} \simeq 0.9$ km/s, where $T \sim 100$ K, that is, for magnetic field values of the order or smaller than $3\,\mu$G, condition $v_A \gg c_s$ is not completely satisfied, and the compression ratio must be smaller than the value estimated by (9.36). In fact, from Eq. (9.36) and taking $u_0 \simeq 10$ km/s, we get $\rho_1/\rho_0 \simeq 120$, and from (9.48) we find $\rho_1/\rho_0 \simeq 7$, although this expression is not completely applicable here. The numerical values obtained are only approximate, but we can observe that the introduction of a magnetic field tends to decrease the compression ratio in the case of an isothermal shock. In this case, part of the shock energy is used to increase the magnetic pressure, as the increase in the density in the post-shock region concentrates the magnetic field lines.

Exercises

9.1. Write the continuity equation in the one-dimensional case starting with Eq. (1.6). Assuming steady state, prove Eq. (9.1).

9.2. Prove Eqs. (9.15)–(9.17).

9.3. (a) Solve simultaneously Eqs. (9.16) and (9.17), and obtain the following relations:

$$\frac{\rho_1}{\rho_0} = \frac{(\gamma + 1)\, M_0^2}{(\gamma - 1)\, M_0^2 + 2}$$

$$\frac{P_1}{P_0} = \frac{2\,\gamma\, M_0^2 - (\gamma - 1)}{(\gamma + 1)}\ .$$

(b) Use result (a) and the perfect gas equation of state and show that

$$\frac{T_1}{T_0} = \frac{[2\,\gamma\, M_0^2 - (\gamma - 1)]\,[(\gamma - 1)\, M_0^2 + 2]}{M_0^2\,(\gamma + 1)^2}\ .$$

9.4. Consider an adiabatic flow where the Rankine-Hugoniot are written in the form

$$\rho\, u = \phi$$

$$P + \rho\, u^2 = \zeta$$

$$\frac{1}{2}\, u^2 + \tfrac{5}{2}\,\frac{P}{\rho} = \xi.$$

By defining a reference velocity $\bar{u} = \zeta/\phi$ and a non-dimensional variable $\eta = u/\bar{u}$, (a) derive expressions for the total energy per unit mass, the internal energy, the kinetic energy, and the Mach number as functions of η. (b) Consider the non-dimensional variable σ, which represents entropy,

$$\sigma = \frac{P\,\rho^{-5/3}\,\phi^{2/3}}{\bar{u}^{8/3}}\ ,$$

and show that $\sigma = \eta^{5/3}\,(1 - \eta)$. (c) Show that, if the total energy is to be conserved and the entropy increases, the internal energy must increase at the expense of the kinetic energy, and the Mach number decreases (see Dyson and Williams 1997, Chap. 6).

9.5. We may define the Mach number in region 1 (*downstream*) in a similar way to Eq. (9.12). Show that

$$M_1^2 = \frac{(\gamma - 1)\, M_0^2 + 2}{2\,\gamma\, M_0^2 - (\gamma - 1)}$$

and that the flow is always subsonic.

Bibliography

Dyson, J., Williams, D.A.: The Physics of the Interstellar Medium. Taylor & Francis, Oxford (1997) (Intermediate level text on the physics of the interstellar medium, including a very accessible analysis of shock waves, especially as an application of the interaction of stars and the interstellar medium. Presents in detail the mathematical formalism leading to the variations of the physical properties in the shock zone. The treatment in Sections 9.5 and 9.6 was partially based on this reference)

Lamers, H.J.G.L.M., Cassinelli, J.P.: Introduction to Stellar Winds. Cambridge University Press, Cambridge (1999) (Referred to in Chapter 1. Includes an analysis of shock waves generated by the interaction of stellar winds with the interstellar medium)

Landau, L., Lifchitz, E.: Mécanique des Fluides. MIR, Moscou (1971) (English edition: Fluid Mechanics. Butterworth-Heinemann, Boston (1995). Referred to in Chapter 1. Contains a chapter with a good discussion of shock waves, jump conditions and applications to perfect fluids)

Maciel, W.J.: Astrophysics of the Interstellar Medium. Springer, New York (2013) (An introduction to the astrophysics of the interstellar medium. Considers some dynamic processes related to the interaction of stars and the interstellar medium, such as stellar winds, supernovae and the expansion of HII regions and planetary nebulae. Translation of a 2002 book originally published in Portuguese (Astrofísica do Meio Interestelar, Edusp, 2002))

Mihalas, D.: Stellar Atmospheres. Freeman, San Francisco (1978) (Referred to in Chapter 1. Presents a rigorous discussion of the hydrodynamic equations for compressible fluids as applied to stellar winds. Derivates the Rankine-Hugoniot conditions for stationary shocks)

Spitzer, L.: Physical Processes in the Interstellar Medium. Wiley, New York (1978) ([Student edition: 1998]. Classic text on the main physical processes in the interstellar medium, in particular the shock waves produces by the interaction of stars and the interstellar medium. The treatment in Section 9.8 is based on this reference)

Zeldovich, Ya.B., Raizer, Yu.P.: Physics of Shock Waves and High Temperature Hydrodynamic Phenomena. Dover, New York (2003) (New printing of a classic advanced text on shock waves, originally published in 1966)

Chapter 10
Stellar Winds: An Overview

Abstract This chapter presents a general view of stellar winds, the main observational evidences and driving mechanisms. An estimate is made of the mass loss rates for hot and cool stars. The chapter ends with some comments on the interaction of the winds with the interstellar medium and the effects of winds on the evolution of the stars.

10.1 Introduction

Mass loss is a common phenomenon in stars, and may occur in a *catastrophic* way, as in supernova explosions, when most of the stellar mass is lost, or more quietly, in a *continuous* way, as in the case of the solar wind. Some observational evidences of these phenomena can be traced back to the sixteenth century, with the observation of a supernova in 1572 by Tycho Brahe, or the first observations of the variable star P Cygni. However, only after the first decades of the twentieth century it was possible to observe and analyze this process in a systematic way. In the case of novae, some images taken after the mass ejection indicated the presence of concentric expanding shells, suggesting that the gas was moving at velocities larger than the escape velocity at the stellar atmosphere, which was an evidence of mass loss.

In this book we are basically interested in the mass loss processes that occur in a *continuous* way, which is a characteristic of *stellar winds*. These processes can be observed in luminous hot stars, such as the blue supergiants, in solar-type stars and in cool giants and supergiants. The main evidences, as we will see in Sect. 10.2, are of spectroscopic nature. In particular, *P Cygni profiles* are observed in many spectral lines, as can be seen in Fig. 10.1, which was obtained from observations of the 10,830 Å He I line. Usually, these profiles show an emission component, displaced towards larger wavelengths relative to the line center λ_0, and an absorption component, displaced towards shorter wavelengths. The profiles can be interpreted assuming that they originate in an expanding envelope around the star, in which

W.J. Maciel, *Hydrodynamics and Stellar Winds: An Introduction*, Undergraduate
Lecture Notes in Physics, DOI 10.1007/978-3-319-04328-9_10,
© Springer International Publishing Switzerland 2014

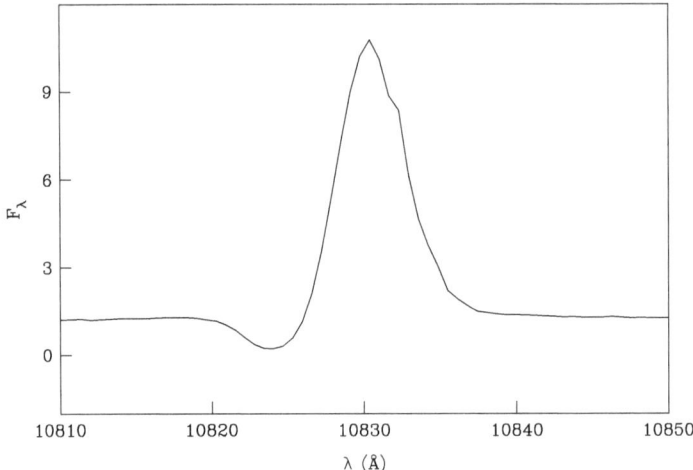

Fig. 10.1 Profile of the He I line at 10,830 Å in the star P Cyg. The stellar flux is in arbitrary units (A. Damineli, IAG)

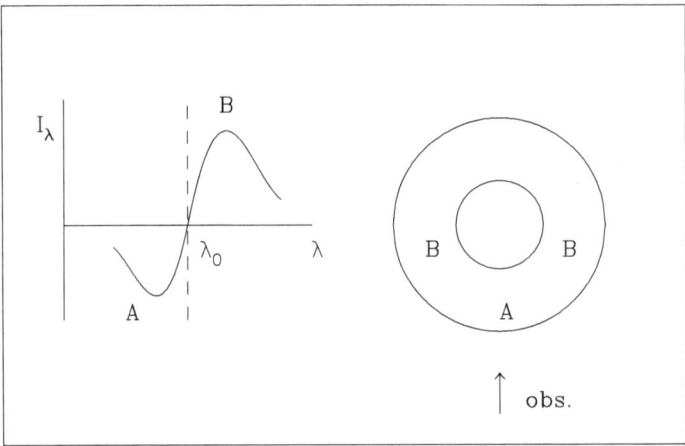

Fig. 10.2 Model for the interpretation of P Cygni profiles. The absorption is produced in region A, while the emission comes from region B

the absorption component is produced in region A and the emission component in region B, as shown in Fig. 10.2. The earliest application of these ideas to stars presenting P Cyg profiles are from the decade of 1920, concerning Wolf-Rayet stars. Since then, the observational evidences and theoretical developments have accumulated in an impressive way both for hot and cool stars, since the mass loss phenomenon appears to be important in all parts of the HR diagram. Research on the solar wind was of particular importance in the early years, and the main results were extended later to other stars.

The study of stellar winds can be approximately divided into four parts: (i) the analysis of the observations and methods used to determine the wind properties; (ii) the study of the mechanisms responsible for the mass ejection; (iii) the study of the effects of the winds on the neighbouring regions of the interstellar medium, and (iv) the study of the effects of the winds on the evolution of the stars. In this chapter, we will present a general overview of these aspects. A more detailed approach can be found in the textbook by Lamers and Cassinelli (1999).

10.2 Observational Evidence

Generally speaking, the observational evidences of stellar winds are used to estimate the mass loss rate $dM/dt = \dot{M}$ ($M_\odot$/year) and the terminal velocity $v = v_f = v_\infty$ (km/s). Low and intermediate mass stars ($M < 10\,M_\odot$) have very small rates on the main sequence, $\dot{M} \leq 10^{-10}\,M_\odot$/year. During their late evolution stages on the giant branch, the rates may reach about $10^{-6}\,M_\odot$/year or higher, as in the superwind that originates planetary nebulae. As a consequence, in these late stages the stellar evolution is significantly affected by mass loss. The ejection velocities in these stars are typically of the order of a few km/s up to tens of km/s. Concerning massive stars ($M > 10\,M_\odot$), even on the main sequence the observed mass loss rates are important, and may be about $10^{-5}\,M_\odot$/year, thus affecting the stellar evolution. The corresponding terminal velocities are much higher, of the order of a few thousands km/s. In the following we will summarize the main observational evidences of stellar winds with some estimates of properties such as the terminal velocity v_f and the rate $\dot{M}$. A detailed description of the methods used to determine the mass loss rate is beyond the goals of this book, and the interested reader may consult the bibliography at the end of the chapter.

10.2.1 Example: P Cygni Profiles

The presence of P Cyg profiles is extremely important in the identification of stellar winds, particularly in the case of resonance ultraviolet lines in hot stars (O, B, Wolf-Rayet stars, and central stars of planetary nebulae). The obtained mass loss rates are typically $\dot{M} \geq 10^{-8}\,M_\odot$/year. These lines are usually formed by *resonant scattering*, in which an atom in the stellar envelope absorbs a photon from the photosphere leading to a photoexcitation followed by deexcitation, and the photon is reemitted practically in the same frequency as the original photon, taking into account the Doppler effect. Many examples of lines with P Cyg profiles are observed, particularly in the ultraviolet (UV) spectrum of hot stars. Table 10.1 shows some typical UV lines.

Table 10.1 Some typical
ultraviolet spectral lines

Ion	λ (Å)
CII	1,334.532
	1,335.708
CIII	1,175.670
CIV	1,548.195
	1,550.770
NIV	1,718.551
NV	1,238.821
	1,242.804
OVI	1,031.928
	1,037.619

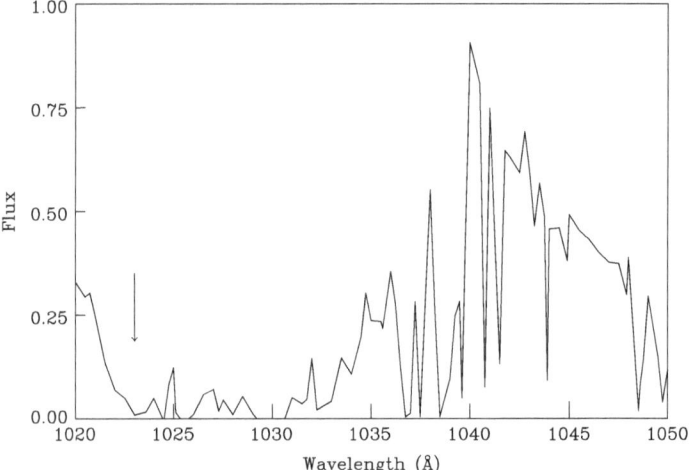

Fig. 10.3 FUSE spectrum of the planetary nebula NGC 7009 showing the P Cyg profile of the OVI line at 1,032/1,039 Å

Depending on the wind column density in region A (Fig. 10.2), an absorption component will be produced, blueshifted by up to 3,000 km/s due to the Doppler effect. For large column densities, $N \geq 10^{15}$ cm^{-2}, the P Cyg profile also presents a redshifted emission component.

10.2.2   Example: Terminal Velocity in the Wind of NGC 7009

Planetary nebulae have very hot central stars (CSPN), so that their ultraviolet spectra often presents strong P Cyg profiles caused by stellar winds. Figure 10.3 shows a tracing of part of the spectrum of the planetary nebula NGC 7009 obtained with the FUSE satellite (Far Ultraviolet Spectroscopic Explorer), where the flux is in arbitrary units. We can observe the strong P Cyg profile of the OVI doublet at

1,032/1,039 Å. We may assume that most of the observed profile is determined by the strongest line in the doublet, with rest wavelength $\lambda_0 = 1,031.928$ Å. In this case, the terminal velocity of the wind can be estimated from the blueshifted absorption component. We notice that the limiting wavelength is $\lambda \simeq 1,023$ Å, as indicated by the arrow in the figure. The corresponding velocity is simply given by

$$v_f \simeq \frac{\Delta\lambda}{\lambda_0} c \simeq \frac{1,023 - 1,031.928}{1,031.928} 3.0 \times 10^{10} \simeq -2,600 \, \text{km/s} \tag{10.1}$$

where the negative sign means that the gas responsible for the absorption is approaching the observer.

In principle, the terminal flow velocity v_f, the velocity law, and possibly the mass loss rate can be determined from high-resolution observations of different ions. Some effects must be considered that limit the use of these lines, such as saturated profiles, turbulence in the line-forming region, ionization conditions in the wind, presence of shocks, absence of local thermodynamic equilibrium (NETL), etc. For example, in the case of non-saturated P Cyg profiles, the radial distribution $n_i(r)$ of the ions responsible for the spectral line can be determined, and the mass loss rate can be written as

$$\dot{M} = f[n_i(r), v(r), A_i, q_i, cc], \tag{10.2}$$

where A_i is the abundance of the considered element relative to H, q_i is the fraction of ions in the excitation and ionization stages corresponding to the spectral line, $v(r)$ is the velocity law, and cc refers to the chemical composition of the wind.

10.2.3 Example: Terminal Velocity and Model Fitting in CSPN

In more detailed models, the terminal velocity is determined not only from the blue-shifted absorption edge, as shown in Fig. 10.3, but also involves the detailed calculation of the line profile. Figure 10.4 shows the CIV line doublet at 1,548.2, 1,550.8 Å in the spectrum of the CSPN NGC 6905. This is a hydrogen-deficient CSPN of [WCE] type, a condition that arises from thermal pulses after the star leaves the Asymptotic Giant Branch (AGB), so that the star has a spectrum resembling those of the Population I Wolf-Rayet stars. The ragged continuous line shows the observed spectra, and if we apply the same simplified procedure of the previous example, we would estimate a limiting wavelength of $\lambda \simeq 1,536$ Å, so that the terminal velocity would be

$$v_f \simeq \frac{\Delta\lambda}{\lambda_0} c \simeq \frac{1,536 - 1,548.2}{1,548.2} 3.0 \times 10^{10} \simeq -2,360 \, \text{km/s} \, .$$

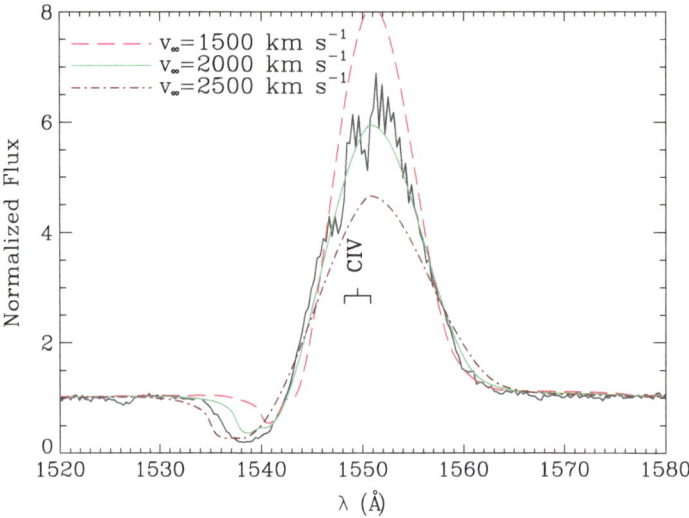

Fig. 10.4 Ultraviolet spectrum of the CSPN NGC 6905 with three numerical fits to the CIV line at 1,548/1,550 Å corresponding to the terminal velocities of 1,500, 2,000, and 2,500 km/s (Keller et al. 2011)

Adopting now the recent grid of synthetic spectra obtained with the state-of-the-art stellar atmosphere numerical code by Keller et al. (2011), we obtain the three additional lines shown in the figure, calculated for terminal velocities of 1,500, 2,000, and 2,500 km/s. We can see that the intermediate velocity, $v_f \simeq 2,000$ km/s (solid continuous line) gives the best fit to the observed spectrum.

10.2.4   Example: Turbulence in Stellar Atmospheres

Spectral lines are not infinitely narrow, as several broadening mechanisms operate in order to define a line profile. The main mechanisms are the natural broadening, a consequence of Heisenberg's uncertainty principle, the Doppler effect due to the motion of the absorbing or emitting particles in the line of sight, and the pressure broadening, caused by the superposition of the energy levels in high density gases. In the atmospheres of stars such as the CSPN NGC 6905 the Doppler effect is dominant, but the gas turbulence due to small scale random motions of the absorbing particles in the atmosphere may also play a role. Although a good fit is obtained from Fig. 10.4, it can be seen that the calculated profile is somewhat narrower than the observed profile, especially at the blue-shifted absorption component. This is corrected in Fig. 10.5, which shows the P Cyg profile of the CIV doublet at 1,548.2, 1,550.8 Å as in the previous figure, where the three additional lines represent the fits for different values of the turbulent velocity from the numerical

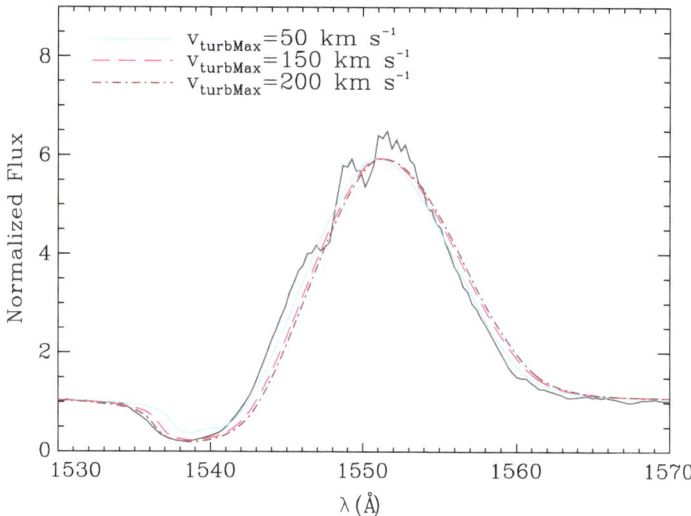

Fig. 10.5 The same as Fig. 10.4 with numerical fits for a terminal velocity of 2,000 km/s and maximum turbulent velocities of 50, 150, and 200 km/s (Keller et al. 2011)

model of Keller et al. (2011). In this case the best fit occurs for turbulent velocity of $v_t \simeq$ 150–200 km/s, and the corrected terminal velocity used in the fit is $v_f \simeq$ 2,000 km/s. The numerical model allows the determination of the main stellar physical parameters, which are $T_* \simeq$ 150,000 K, $dM/dt \simeq 10^{-7} \, M_\odot/\text{year}$, and $v_f \simeq$ 2,000 km/s.

10.2.5   Example: The Hα Emission Line

Hot stars (O, B, Wolf-Rayet) with mass loss rates $\dot{M} \geq 10^{-7} \, M_\odot/\text{year}$ may show the hydrogen Hα line ($\lambda 6{,}563$ Å) in emission, and the corresponding profile is a good indicator of mass loss. The line is formed by the collision of a wind ion with an electron, and the recombination process leaves the ion in an excited state, followed by decay down to the fundamental state, when several emission lines are produced. Apart from Hα, other lines may also be observed, such as the Paschen and Bracket series He lines in the infrared part of the spectrum, other He lines, such as HeI $\lambda 10{,}830$ Å (Fig. 10.1) and HeII $\lambda 1{,}640$ and 4,686 Å. As in the case of P Cyg profiles, the profiles of these lines may give some information on the mass loss rate and the wind velocity law. The profiles are usually symmetric and the emission extends to velocities smaller than the terminal velocity, since the lines are formed in denser regions, where the wind is still in the acceleration phase.

The theoretical treatment of line formation in moving media is due to the classic work by Sobolev (1960), which includes a model for the radiative transfer defined in terms of the photon escape probability. Some complicating effects must be taken into account, such as the radiation optical depth in the line, the need of the distance to the star, and the geometric dilution effect. Knowing the velocity law and the temperature profile in the wind, the mass loss rate can be written in terms of the line luminosity L. For optically thin lines, it can be shown that

$$\dot{M} \propto v_f \sqrt{L} \,. \tag{10.3}$$

Therefore, if the terminal velocity can be obtained from the P Cyg profile, then Eq. (10.3) can be used to estimate the mass loss rate.

10.2.6   Example: Infrared and Radio Excess

In hot stars with ionized winds an excess of radio and infrared radiation is observed, caused by free-free (Bremsstrahlung) emission by the wind. The excess radiation is measured relative to the stellar Planck emission, which corresponds approximately to the Rayleigh-Jeans distribution, which is valid for long wavelengths. Measurements of the energy distribution in this spectral region provide an useful method to determine the properties of the stellar winds. This method can be applied to Be stars, particularly for infrared radiation at $\lambda < 10\,\mu m$. In radio wavelengths the method allows the determination of mass loss rates higher than $10^{-6}\,M_\odot/year$ in nearby hot stars (O, B, Wolf-Rayet).

In this case, the stellar mass loss rate can be determined from the velocity law and gas density, which is obtained from the energy distribution of the excess radiation and the temperature profile. The luminosity of the free-free radiation varies approximately as $v^{0.6}$, where v is the radiation frequency. Naturally, this refers to continuum radiation, and not to spectral lines, so that no information can be obtained on the gas expansion velocity by the Doppler effect, in which case other sources of information must be used.

An application of the free-free emission theory shows that the mass loss rate depends on the monochromatic radio flux f_v, but it is practically independent of the wind temperature, so that it can be written as

$$\dot{M} \propto v_f \, (f_v \, d^2)^{3/4} \, v^{-1/2} \,, \tag{10.4}$$

where d is the distance to the star.

10.2.7   Example: Molecular Emission of Cool Stars

Slow massive winds, with velocities $v_f \simeq 10\,km/s$ and rates $\dot{M} \sim 10^{-6}\,M_\odot/year$ can be identified from observations of rotation transitions of CO molecules in the envelopes of cool giants. Depending on the wind conditions, including the fraction of atoms or molecules in the upper energy levels and the velocity gradient in the photon propagation direction, *maser* amplification of the emitted radiation may also occur.

Examples of a few molecular lines showing evidence of mass loss are $\lambda 1.3$ mm ($\nu = 230.7$ GHz) and $\lambda 2.6$ mm ($\nu = 115.4$ GHz) for the CO molecule; $\lambda 3.5$ mm and 6.9 mm (SiO) and $\nu = 1{,}612, 1{,}665, 1{,}667$ MHz ($\lambda \simeq 18$ cm) for OH. The envelopes are extended and relatively dense, so that the mass loss rate is large, even if the terminal velocity is low. The most important case is that of the CO molecule, which is extremely stable and abundant in the atmospheres of oxygen-rich stars (O/C > 1, where O and C refer to the oxygen and carbon abundances by number of atoms), or carbon-rich stars (O/C < 1).

Analogously to the previous case, the luminosity of the spectral line can be related to the mass loss rate, so that the latter can be inferred. Different results are obtained for optically thin or thick lines, which can be usually determined from the observed profiles. For example, for optically thick CO emission lines we have

$$\dot{M} \propto T_b \, v^2 \, d^2 \, f_{CO}^{-0.85} \,, \tag{10.5}$$

where T_b is the antenna temperature at the line center and f_{CO} is the CO abundance relative to the H_2 molecule. Equation (10.5) can be used to estimate the mass loss rate in asymptotic giant branch stars (AGB stars).

10.2.8 Example: Grain Infrared and Millimeter Emission

Cool giants present infrared and millimeter radiation excesses due to the solid grains embedded in the expanding envelope. Observations of the energy distribution allow the estimate of mass loss rates of the order of 10^{-6}–$10^{-7}\ M_\odot$/year.

The dust energy distribution can be observed in the infrared and millimeter spectral regions, and are clearly distinguished from the free-free emission mentioned before. Depending on the grain column density, the observed spectrum may include a stellar component, apart from the Planck emission by the dust. Analysis of this emission includes the study of the space distribution of the dust temperature, emissivity, energy distribution and physical properties of the circumstellar grains, which are poorly known. Measurements of the polarization caused by the grains can be associated to the observations of the infrared excess, in order to determine the grain density in the envelope. As in the case of the free-free radiation, we are dealing here with continuous radiation, so that any information on the velocity structure must be obtained by other methods.

The luminosity of the infrared emission can be obtained in terms of the mass loss rate, depending on the emission optical depth. For example, in the case of optically thick winds, the mass loss rate can be written as

$$\dot{M} \propto \mu \, v \, d^2 \, L_*^{-1/2} \, F \, \lambda^{1/2} \, \kappa^{-1} \,, \tag{10.6}$$

where μ is the gas-dust ratio, v is the wind velocity, d is the distance to the star, L_* is the stellar luminosity, F is the IRAS (*Infrared Astronomical Satellite*) flux in the $60\,\mu$m spectral region, λ is the average wavelength of the circumstellar emission, and κ is the dust opacity. In this case, we can estimate the dust mass loss rate $\dot{M}_d$ and the gas mass loss rate, which we call simply $\dot{M}$, since $\dot{M} \gg \dot{M}_d$.

10.3 Isothermal and Non-isothermal Winds

The gas temperature plays an important role in the study of stellar winds, since it directly reflects the energy gain and loss processes which are included in the energy equation. The simplest case is that of *isothermal winds*, in which the temperature is constant throughout the envelope. In this case, the energy equation is very simple, and can be written as $T =$ constant, as we have seen in Chap. 4 (Eq. 4.1).

The solution of the hydrodynamic equations is particularly simple assuming that the momentum equation includes only the gravitational and the pressure gradient terms. According to the results of Sect. 8.5, the system of equations has only one critical solution in which the wind goes from the subsonic regime to the supersonic regime, passing through the point where $v = c_s$ at a certain critical radius r_c (see Sect. 8.4). In this case, the mass loss rate can be estimated for example from the values of r, ρ, and v at the critical point. This model is extremely idealized, and can be improved by the addition of a force proportional to r^{-2} in the whole or parts of the envelope. This force may be radiative, as in the winds of hot stars driven by the stellar radiation pressure on spectral lines of abundant ions, or in the winds of cool stars under the action of the radiation pressure on molecular lines or dust grains. Another example is a force proportional to $v\,(dv/dr)$, as in hot stars under the action of the radiation pressure due to optically thick spectral lines (see Exercise 10.3).

In the simplest isothermal models, the wind structure below the sonic point is basically defined by the hydrostatic equilibrium condition, while in the supersonic region it is defined by the forces producing the acceleration up to the terminal velocity. The model can be generalized by considering temperature gradients, which leads to changes in the wind structure, since (i) the Euler equation must include new components of the pressure gradient, which alter the velocity profile, and (ii) there are changes in the sound speed and the Mach number.

A stellar wind can be associated to a gravitationally bound atmosphere, in which the potential energy per unit mass is $-G\,M_*/R_*$. The wind may escape from the potential well represented by the star if its energy is positive at large distances from R_*, of the order of $v_f^2/2$, that is, some physical process responsible for the temperature variation must provide the wind energy and/or momentum. Depending on the energy or momentum transfer processes, profound changes may then occur in the velocity structure and in the mass loss rate relative to the isothermal model. The treatment of the hydrodynamic equations is more complicated, as it must reflect the different physical processes involved, and the solution may include several critical points.

The energy transferred to the gas in the wind may be in the form of heat or momentum, or both. For example, the dissipation of acoustic waves at the base of the solar corona corresponds to the addition of heat, while the action of the radiation pressure on spectral lines of abundant ions in hot stars corresponds to momentum transfer, that is, energy is added by the work done by the external force. Both processes must be considered in the energy equation and/or in the Euler equation, but their effect on the wind structure is different. For example, while momentum transfer to the gas by radiation corresponds to an external force in opposition to the gravity force, the addition of heat tends to increase the gas temperature, decreasing the temperature gradient and thus the pressure gradient, so that the pressure forces have a new component in the same direction as the stellar gravity. In principle, both energy and momentum transfer processes are needed in order to obtain transonic solutions for the wind equations, that is, solutions that are subsonic in the acceleration region and supersonic in the external parts of the envelope. Apart from that, parameters such as the mass loss rate and the terminal velocity depend on the exact location of the region where energy is transferred to the gas, and on the way the external force on the expanding gas depends on position or velocity.

As in the study of stellar structure, the adoption of *polytropic winds* allows the inclusion of temperature gradients, keeping the solution of the hydrodynamic equations relatively simple. In this case, a polytropic index Γ is defined such that $d \ln P / d \ln \rho = \Gamma$. The case $\Gamma = 1$ corresponds to isothermal winds, and $\Gamma = 5/3$ corresponds to adiabatic winds. Γ is related to the specific heats ratio γ, and should not be confused with the Γ parameter defined by Eq. (2.75). Intermediate values of this parameter produce approximate solutions to real winds. Depending on the value of the polytropic index, the gas in the envelope may or may not escape from the star. Also in this case, the mass loss rate can be estimated from the conditions in the subsonic region, and the wind structure depends critically on the polytropic index.

10.3.1 Example: Isothermal Wind with an External Force Proportional to r^{-2}

Winds driven by optically thin lines or by electron scattering have an additional force inversely proportional to the distance squared, namely $F \simeq A\, r^{-2}$, where A depends on the wind parameters. Assuming again isothermal winds, the Euler equation is again (8.23), with an additional term $+A/r^2$ in the second member. With the same procedure of Sect. 8.4, that is, using Eqs. (8.24)–(8.27), it is easy to show that we obtain an equation like (8.40), with the only difference that now the numerator of the second member has the additional term $+A/r^2$. Therefore, the topology again includes one critical solution for a given set of wind parameters. The actual solution depends on whether $A \neq 0$ throughout the wind or just in parts of it.

For instance, if $A \neq 0$ from a certain distance r_d onwards, as in the case of grain formation in the outer atmospheres of cool stars, the effect on the terminal velocity v_f and mass loss rate dM/dt will depend on the position of r_d. If $r_d > r_c$, the terminal velocity will be increased, but the mass lass rate is unaffected, as the wind structure in the subsonic part of the flow is the same; if $r_d < r_c$, the transferred wind momentum is larger in the subsonic part of the flow, so that the mass loss rate may be larger than in the case of no external force apart from the pressure forces. For details on this condition as well as in the case of a force proportional to $v \, (dv/dr)$ see Lamers and Cassinelli (1999), Chap. 3.

10.4   Mechanisms Responsible for the Ejection

Many different mechanisms have been proposed to explain the mass ejection observed on the HR diagram. A detailed discussion of these processes can be found in the book by Lamers and Cassinelli (1999) and references therein. In this section we will present an introduction to the main models and their applications to different stellar types.

10.4.1   Example: Coronal Winds

Cool stars on the main sequence with spectral types F5 or later and some giant stars probably have chromospheres and coronae as in the Sun. Coronal winds are produced by the gas pressure in this high temperature region, reaching $T \sim 2 \times 10^6$ K in the case of the Sun. Such temperatures may be produced by dissipation of mechanic energy or by reconexion of magnetic fields in the convective subphotospheric layers of these stars.

The momentum equation for coronal winds is relatively simple, including only gravity and gas pressure. Thermal conduction is important at these high temperatures, and should be included in the energy equation. Conduction transfers energy from the high temperature regions to the low temperature regions, so that the temperature gradient tends to decrease in the regions where conduction is efficient.

The system of equations is usually solved numerically, and the solution must adjust to the physical conditions in the neighbourhood of the critical point, where the gas velocity approaches the sound velocity. In the hot, dense regions the flow has low velocities. As the gas cools, heat conduction from the hotter layers to the cooler layers lead to transonic solutions, reaching the velocities observed in the solar wind. Typical velocity profiles show that the gas can be accelerated to about 300 km/s within 50 stellar radii, where the temperature decreases by a factor of 4. The expected mass loss rates are low, of the order of 10^{-14} $M_\odot$/year, as we have seen in the case of the Sun. The estimated rates are in the approximate range of $4 \times 10^{-16} < \dot{M} \, (M_\odot/\text{year}) < 10^{-13}$.

Generally speaking, the structure of coronal winds and the corresponding mass loss rates are similar to the isothermal and polytropic winds, since the physical processes included in the Euler equation (pressure forces and gravity) are the same. However, the observed terminal velocities are smaller than predicted by these idealized models, since they assume that energy is also transferred in the supersonic region, while in realistic models this assumption is usually not adopted.

10.4.2 Example: Winds Driven by Sound Waves

Convective motions in the stellar subphotospheric regions may generate *acoustic waves* that propagate through the atmosphere. These waves are associated with an additional pressure component, the *wave pressure* $P_o \sim (1/2)\,\rho\,v_o^2$, where v_0^2 is the mean quadratic velocity of the oscillations, and the corresponding pressure gradient term can in principle produce a force opposing gravity, which may lead to a stellar wind.

For a particle moving in an oscillating field, if the amplitude of the oscillations decreases with distance, part of the kinetic energy of the oscillations can be transferred to the expanding gas, accelerating the particle.

The simplest models assume isothermal winds driven by sound waves with no dissipation of the acoustic energy; the mass loss rates are usually small and can be obtained from the density, position and velocity in the subsonic region, depending on the amplitude of the sound wave. On the other hand, for higher amplitudes, comparable to the sound speed, dissipation of acoustic energy must be taken into account, affecting the mass loss rate, which is a function of the luminosity of the acoustic wave. This process may be important in cool stars with low surface gravity, such as the AGB stars, as long as the winds are approximately isothermal or if the driving mechanism is coupled with the action of the stellar radiation pressure on grains in the supersonic region.

10.4.3 Example: Winds Driven by Circumstellar Dust

Solid grains can be formed in the atmospheres and envelopes of cool giant stars, and may be ejected by the action of the stellar radiation pressure, leading to large mass loss rates. This process is based on the momentum transfer from the radiation field to the grains, and from the grains to the gas. The grains can absorb stellar radiation in a large range of wavelengths, so that the dust driven winds are basically driven by continuous radiation. In this respect, they differ from the winds in hot stars, driven by the action of radiation pressure on lines formed by resonant scattering, or from cool stars under the action of radiation pressure on molecular lines and bands. For continuum radiation driven winds, the Doppler shifts are not important.

For example, in a cool star with $v_f \simeq 30$ km/s, radiation at $\lambda_0 = 1\,\mu$m will have a maximum shift given by $\Delta\lambda = \lambda_0(v_f/c) \simeq 1$ Å. In this small wavelength range the radiation field and the gas opacity can be considered as essentially constant. A different situation occurs in the spectral line driven winds. For example, in a hot star with $v_f \simeq 3{,}000$ km/s, a spectral line centered at $\lambda_0 = 1{,}000$ Å has a Doppler shift up to $\Delta\lambda = \lambda_0(v_f/c) \simeq 10$ Å. In this range the gas opacity may change drastically, since the absorption coefficient varies strongly with wavelength within a spectral line.

For this mechanism to be efficient, a combination is necessary between the envelope scale height, the gas density ρ, and the temperature T, since the grains are only able to form if ρ is sufficiently high at a distance from the star where the effect of the radiation pressure is comparable or larger than the stellar gravity, and the temperature is below the grain condensation temperature. The grains absorb the stellar radiation and emit as black bodies in the infrared part of the spectrum, and their temperature is determined by the equilibrium between these processes.

Oxygen-rich stars, where O/C > 1, are able to form silicate grains such as Mg_2SiO_4, while carbon-rich stars, where O/C < 1, may form SiC or graphite grains. The grain properties are usually not well known, particularly the opacities and efficiency factors for radiation pressure, which are needed in the radiative term of the momentum conservation equation. In pulsating stars, pulsation itself may affect the mass ejection process or even initiate the ejection in the photospheric regions or at the base of the circumstellar envelope by increasing the scale height in the subsonic part of the wind.

Dust-driven winds have a complex structure, particularly in view of the variation of the density and velocity law with position, and also by the existence of a "drift" velocity of the grains relative to the gas. The β law (2.92) is a good approximation with $\beta \simeq 0.5$. Results suggest condensation radii of the order of a few stellar radii in stars with effective temperatures typically in the range of 2,000–3,000 K. Moreover, the mechanism can be used as a complement to another driving mechanism, such as pulsation, radiation pressure on molecular lines, or Alfvén waves.

The mass loss rates can be derived considering that the velocity law must go through the critical point, where in principle we have $v \simeq c_s$. The obtained values are of the order of 10^{-5}–$10^{-7}\,M_\odot$/year, in good agreement with the observational determinations. The mass loss rate ($M_\odot$/year) in cool stars, particularly red supergiants, can be approximated by an empirical formula:

$$\dot{M} \simeq 4 \times 10^{-13}\,\eta\,\frac{(L_*/L_\odot)\,(R_*/R_\odot)}{M_*/M_\odot}\,, \tag{10.7}$$

where the stellar luminosity, radius and mass are given in terms of the solar values, and η is an adjustable parameter, typically in the range $0.3 \le \eta \le 3$. This is *Reimers formula*, and was obtained using a sample of red supergiants in binary systems for which there are relatively accurate determinations of dM/dt, M_*, R_* and L_*. The simplest quantity having dimensions of mass/time that can be formed from the basic stellar parameters, M_*, R_*, and L_* is $L_*/g\,R_*$, which is equivalent to $L_*\,R_*/M_* = L_*/(M_*/R_*)$, since $g \propto M_*/R_*^2$. From Eq. (2.93)

$g R_* \propto M_*/R_* \propto v_e^2$; if the mass loss rate is proportional to this quantity in different stars, the same fraction of the stellar luminosity is used to provide the potential energy per unit mass of the material escaping from the star.

We can have an idea of the mass loss rate in a radiative wind assuming that all stellar photons are absorbed once, which is the so-called "single scattering limit". In this case, the wind momentum per unit time is $\dot{M} v_f$, which must be equal to the stellar radiative momentum per unit time, given by L_*/c. We have then

$$\dot{M} \simeq \frac{L_*}{v_f \, c} . \tag{10.8}$$

In the case of a single scattering, (10.8) is an upper limit, since not all stellar photons are in fact absorbed. However, the same photon can be absorbed and reemitted or scattered several times, which may increase the mass loss rate by a considerable factor, so that approximately correct rates can be obtained with Eq. (10.8). Detailed models show that, in a better approximation, this equation must be multiplied by τ, the wind optical depth in the supersonic region, namely

$$\dot{M} \simeq \frac{L_*}{v_f \, c} \tau . \tag{10.9}$$

The optical depth is a quantity that may reach values much higher than unity, but an upper limit can be estimated by considering that the total wind kinetic energy per unit time is limited by the energy radiated by the star, which is characterized by its luminosity L_*. In this case we have

$$\frac{1}{2} \dot{M} v_f^2 \leq L_* , \tag{10.10}$$

from which we derive an upper limit for the optical depth given by

$$\tau \leq \frac{2 c}{v_f} . \tag{10.11}$$

Since the terminal velocity is of the order of a few thousands km/s in the fastest winds, it can be seen that the optical depth upper limit is a very large number.

Figure 10.6 shows a comparison between the results given by (10.7) (solid lines) and (10.8) (dashed lines) adopting typical conditions for the winds in cool giants and supergiants. Reimers formula (10.7) was used with $\eta = 1$, and the plot shows $\dot{M}$ as a function of the luminosity for some typical values of the ratio M_*/R_* given in solar units, that is, $(M_*/M_\odot)/(R_*/R_\odot)$. The limit (10.8) is calculated for terminal velocities in the range $10 < v_f \,(\text{km/s}) < 200$. We can see that the observed values, of the order of 10^{-5}–$10^{-7} \, M_\odot/\text{year}$, can be reproduced by both equations for velocities and mass/radius relations that are typical of cool giant and supergiant stars, showing that these equations are correct, at least as a first approximation.

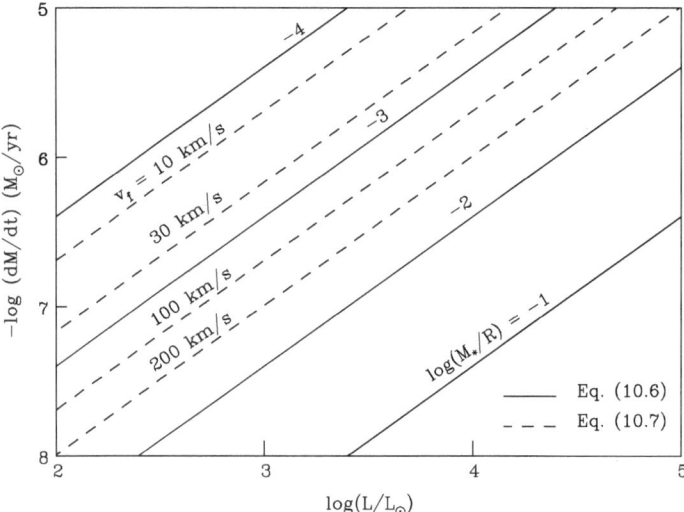

Fig. 10.6 Mass loss rates in cool stars using Eqs. (10.7) (*solid lines*) and (10.8) (*dashed lines*) as functions of the stellar luminosity

In stars where pulsation is the only driving mechanism, the most external layers where the pulsation mechanism is efficient are displaced outwards up to a certain amplitude, falling back to the original position without mass ejection. Grain condensation followed by the action of radiation pressure increases the acceleration of the pulsating layers, which are effectively ejected by the star, so that a true mass loss process occurs. For these stars, the mass loss rates can be estimated as a function of the stellar and dust parameters, and also by some empirical correlations, such as the equation

$$\log \dot{M} \simeq -11.4 + 0.0123 \, P \, , \tag{10.12}$$

where $\dot{M}$ is in $M_\odot$/year and the stellar pulsation period P is in days. This relation is valid for stars with masses of about $2.5 \, M_\odot$ and periods lower than 600 days, such as Mira variables, AGB stars and OH/IR stars. For example, for a Mira variable with a period $P = 400$ days, we estimate a mass loss rate $dM/dt \simeq 3.3 \times 10^{-7} \, M_\odot$/year.

10.4.4   Example: Two-Fluid Hydrodynamics – AGB Stars

AGB (from Asymptotic Giant Branch) stars are cool giant stars that have moved away from the main sequence and have typically adjacent He and H nuclear burning shells. They present strong winds with typical mass loss rates of 10^{-7}–$10^{-6} \, M_\odot$/year and terminal velocities of 10–20 km/s driven by radiation pressure on dust grains. The corresponding wind equations are typical of two-fluid hydrodynamics, that is, the main hydrodynamics equations (continuity, momentum conservation and energy) must be written both for gas and dust. Naturally, these equations are coupled, in particular concerning the source terms in the continuity equation and the drag force term characteristic of grain-gas interaction in the Euler equation. The source terms must include the gas that condenses into dust and the inverse process, in which evaporation removes particles from the grains. Grain processes include nucleation, growth and evaporation, in which matter is transferred from one state (gas or dust) to another. In the Euler equation the main modification is the addition of the drag force, since the grain pressure due to thermal motions is usually negligible. The radiation pressure term on the gas may be important, particularly if molecular lines and bands are taken into account. The grain drag force depends on grain size, which introduces a further complication, since a grain size distribution is more likely than the possibility of identical grains (for details see Habing and Olofsson 2004).

10.4.5   Example: Winds Driven by Line Radiation

This is the most important mechanism for the winds of hot stars with O, B, spectral types and luminosity classes V, III, and I, central stars of planetary nebulae and Wolf-Rayet stars. These stars present a large number of intense absorption lines, particularly in the ultraviolet region of the spectrum, where the opacities may be a million times higher than in the continuum. The momentum of the absorbed photons is transferred to the gas in the stellar atmospheres, leading to the ejection of the surface layers. The process is favoured by the Doppler effect due to the velocity gradient, which permits the absorption of unattenuated photospheric photons in different wavelengths compared to the central wavelength.

The ejection process takes into account the momentum and energy transfer from the internal regions of the star to the expanding envelope by absorption and scattering, which includes the possibility of multiple scattering. The transfer is basically due to ions of C, N, O, Ne, S, P, and iron group elements. These ions transfer the acquired momentum by collisions with protons, electrons and He ions, a process called *coulomb coupling*. According to the original work by Lucy and Solomon (1970), the necessary conditions for the transfer require that the ion deceleration time by collisions with the gas particles must be short compared with the time needed to acquire a large drift velocity relative to the gas, In the process of mass loss by the action of radiation pressure on optically thick lines, the mass loss rate can be written approximately as

$$\dot{M} \simeq N \, \frac{L_*}{c^2} \,, \tag{10.13}$$

where N is the effective number of optically thick absorption lines and L_* is the stellar luminosity. As an example, for a star with $L_*/L_\odot \sim 10^5$, we need $N \simeq 150$ in order to get $\dot{M} \sim 10^{-6}\ M_\odot$/year, of the order of the observed values. A better approximation is obtained assuming that all stellar photons are absorbed once by a large number of strong lines, essentially the single scattering limit given by (10.8). With $L_*/L_\odot \sim 10^5$ and $v_f \sim 10^3$ km/s, we get $\dot{M} \sim 10^{-6}\ M_\odot$/year. The photons are scattered almost isotropically, losing their efficiency in the momentum transfer process in the radial direction after the first scattering event, so that Eq. (10.8) is a good upper limit for the mass loss rate of the majority of hot stars, with the probable exception of the Wolf-Rayet stars, which have higher rates for their luminosities.

Alternatively, the mass loss rate can be estimated using an expression such as

$$\dot{M}\, v_f\, \sqrt{R_*} \simeq f(L_*) \tag{10.14}$$

where R_* is the stellar radius. According to some recent work, there is an approximately linear relation between the logarithm of these quantities (see Sect. 10.4.8 and Exercise 10.1). In the general case, the coupled equations must be solved simultaneously in order to determine the mass loss rate, the velocity law and the terminal velocity. Approximations of the velocity law are frequently used, as for example the laws shown in Chap. 2.

The ejection mechanism by the radiation pressure can also be applied to cool stars, in which the absorption is due to intense H_2O or CO molecular lines. Rates up to $10^{-7}\ M_\odot$/year can be obtained, so that the mechanism may work as a complement to the ejection process by circumstellar grains.

A particularly important aspect of the line radiative winds is the radiative transfer, since the stellar photons are absorbed by an expanding gas. The gas expands according to a velocity law $v(r)$, so that the photons that can be absorbed in a given spectral line must be located in a certain "line interaction region", which is basically defined by the line width and the gas velocity, taking into account the Doppler effect. This problem does not have a known rigorous solution, and a probabilistic transfer equation is normally used coupled to an approximation introduced by Sobolev (1960). In this case, it is assumed that the interaction region where the photospheric photons can be absorbed is infinitely thin, so that the line profile is essentially a δ function. As a consequence, the determination of properties such as the optical depth depends only on the local conditions of the absorption region. In practice, the wind particles have thermal and turbulent motions, the line profile is not infinitely narrow, and the interaction region has finite dimensions, characterized by the so-called "Sobolev length". This approximation is not always valid, particularly if the wind density and velocity gradient variations are not negligible within a region with dimensions of the order of the Sobolev length.

Adopting the Sobolev approximation the problem is considerably simplified, since the radiation flux responsible for the gas acceleration depends only on the emitted stellar flux and on the local conditions in the acceleration region. The determination of the radiative acceleration due to the spectral lines involves the "penetration probability", that is, the probability that radiation from the stellar photosphere reaches the Sobolev region, and the escape probability of this region. From the work by Castor, Abbott and Klein (1975), the star can be assumed as a point source, with a posterior correction in view of its finite size.

Realistic estimates of the line radiative forces can be relatively complex, since a large number of lines may contribute significantly to the gas acceleration, which requires the determination of the excitation and ionization conditions of many energy levels from many chemical elements. The total line radiation acceleration can be written in terms of the electron scattering acceleration (see Example 2.5.6), introducing a scale factor known as the "force multiplier", which may reach values up to about 10^4. The solutions allow the determination of the mass loss rate, the density and velocity profiles, which may be approximated by a β velocity law (see Eq. 2.92). The topology of the solutions is qualitatively similar to the isothermal case (see Fig. 8.2), and there is a unique transonic solution that goes smoothly from the subsonic to the supersonic regions.

Also in the case of radiative winds in hot stars the mass loss rate can be significantly increased by multiple scattering, especially if there are many closely packed lines, so that photons with increasing wavelengths can be absorbed as they escape from the star.

Finally, instabilities may occur in the radiative winds, caused by the growth of perturbations in the gas velocity and density and the presence of clumps. The instabilities affect the detailed structure of the velocity and density profiles, although the average values of these quantities are preserved. They are also responsible for some observed phenomena in these stars, such as X-ray emission and the presence of highly ionized ions, known as superionization.

10.4.6 Example: The Velocity Law and the Mass Loss Rate with the Finite Disk Correction

An approximate velocity law for line driven winds in OB stars taking into account the finite disk correction factor can be written as

$$\frac{v(r)}{v_f} \simeq \left[1 - \frac{0.9983 \, R_*}{r} \right]^{0.83}. \tag{10.15}$$

Let us consider a typical O4 star with effective temperature $T_{eff} = 40{,}000\,\text{K}$ and radius $R_* = 15 \, R_\odot$, where $R_\odot = 6.96 \times 10^{10}$ cm is the solar radius. Assuming that the base of the wind is located at $r/R_* \simeq 1$, and that the initial velocity is half the average thermal velocity v_{th} of the protons in the stellar atmosphere, we have

$$v_0 \simeq \frac{v_{th}}{2} \simeq \frac{1}{2} \sqrt{\frac{2 \, k \, T_{eff}}{m_p}} = 1.29 \times 10^6 \, \text{cm/s} = 12.9 \, \text{km/s} \tag{10.16}$$

where $k = 1.38 \times 10^{-16}$ erg/K is the Boltzmann constant and $m_p = 1.67 \times 10^{-24}$ g is the proton mass. The terminal velocity is then

$$v_f \simeq v_0 \,[1 - 0.9983]^{-0.83} = 2.57 \times 10^8 \,\text{cm/s} = 2{,}570 \,\text{km/s} \,. \tag{10.17}$$

In this model, the mass loss rate scales approximately with the stellar luminosity as $\log \dot{M} \propto \log L_*^{1.66}$, and an average fit can be written as

$$\log \frac{dM}{dt} \simeq 1.66 \log \frac{L_*}{L_\odot} - 15 \,, \tag{10.18}$$

where dM/dt is in solar masses per year and $L_\odot = 3.85 \times 10^{33}$ erg/s is the solar luminosity. For spherical stars, the luminosity is given by

$$L_* = 4\,\pi\,R_*^2\,\sigma\,T_{eff}^4 = 1.99 \times 10^{39} \,\text{erg/s} \,, \tag{10.19}$$

where $\sigma = 5.67 \times 10^{-5}$ erg cm^{-2} s^{-1} K^{-4} is the Stefan-Boltzmann constant, so that we have $\log(L_*/L_\odot) \simeq 5.71$, and the mass loss rate is $\log(dM/dt) \simeq -5.52$ or $dM/dt \simeq 3.0 \times 10^{-6} \, M_\odot/\text{year}$ (for details see Lamers and Cassinelli 1999, Chap. 8).

10.4.7   Example: Winds and Metallicities in Hot Stars

The line driven mechanism for OB stars suggests that the mass loss rate should be a function of the stellar metallicity, since the strongest lines in these stars are "metallic" lines from elements such as Fe. The determination of the mass loss rate as a function of the metallicity is a difficult problem, but some empirical relations obtained so far suggest that $\dot{M} \propto Z^m$, where Z is the atmosphere metallicity by mass and m is an exponent to be determined. Some recent work on OB stars in the Milky Way and the Magellanic Clouds suggest that this exponent is typically in the range $0.6 \le m \le 0.9$, with an average value $m \simeq 0.7$. Similar values seem to be appropriate also for Wolf-Rayet stars in the metallicity range $10^{-3} \le Z/Z_\odot \le 1$, where $Z_\odot \simeq 0.02$ is the solar metallicity. However, there are several complications in the determination of this parameter, such as the possible values of the β exponent of the velocity law and the presence of clumping in the stellar winds.

10.4.8   Example: Winds and the Metallicity of CSPN

For a stellar wind with mass loss rate $\dot{M} = dM/dt$ and terminal velocity v_f in a star of radius R_* we can define a quantity called "the modified wind momentum" p_w by

$$p_w = \dot{M}\,v_f\,\sqrt{R_*/R_\odot} \tag{10.20}$$

with cgs units g cm s^{-2}. There is an empirical correlation between p_w and the stellar luminosity (see Exercise 10.1), that can be applied to hot stars in general and to the central stars of planetary nebulae (CSPN), which often display strong winds with terminal velocities up to a few thousands km/s. A comparison of the relation above with data from actual stars shows some scattering that can be at least partially attributed to the effect of the stellar metallicity, since the winds are believed to be driven by line absorption of metallic ions. A recent analysis of this relation suggests that

$$\log p_w \simeq 23.98 + 1.507 \, \log \frac{L_*}{L_\odot} + [\epsilon(O) - 12] \qquad (10.21)$$

(see Maciel et al. 2008) where p_w is in cgs units, $L_*/L_\odot$ is the stellar luminosity in terms of the solar luminosity, and $\epsilon(O)$ is the nebular oxygen abundance relative to hydrogen, measured by number of atoms, that is

$$\epsilon(O) = \log \frac{n_O}{n_H} + 12 \qquad (10.22)$$

where n_O and n_H are the oxygen and hydrogen particle densities, respectively. Let us apply the correlation above to the star He2-131, for which we have $T_{eff} \simeq 30{,}000$ K, $R_* \simeq 5.5 \, R_\odot$, $\log L_*/L_\odot \simeq 4.3$ and $v_f \simeq 500$ km/s. The average oxygen abundance is $\epsilon(O) \simeq 8.67$, assumed representative of the stellar metallicity. We obtain the modified wind momentum $\log p_w \simeq 27.13$, or a mass loss rate $dM/dt \simeq 1.8 \times 10^{-8} \, M_\odot$/year, similar to the observed values for this star.

10.4.9   Example: Winds and Magnetic Fields – Alfvén Waves

Stars usually have magnetic fields, which may be left over from the protostellar cloud or developed later by some dynamo process. The structure of the field is not well known, with the exception of the Sun, by it is very likely that the field has some measurable effects on the stellar winds, particularly for fast rotating stars.

The coupling of the magnetic field with the stellar rotation plays an important role in the angular momentum loss, as in the case of protostars, pre-main sequence stars and early main sequence stars. These processes occur through magnetic rotation winds, and the basic mechanism was originally suggested for the Sun. Apart from the angular momentum loss, which corresponds to a variation in the equatorial velocity from a few hundreds of km/s to about 2 km/s in the case of the Sun, the mass loss rates can also be affected, increasing by an order of magnitude, particularly if the star has an open magnetic field in the fast rotating equatorial region.

A probably more important mechanism may occur in the presence of magnetic fields even in slow rotating stars. In this case, *Alfvén waves* generated at the base of the wind may propagate through the envelope, and the energy and momentum dissipation associated with the waves may lead to envelope acceleration and mass ejection. Together with the processes involving radiation pressure, Alfvén waves are probably the main wind generation mechanism in stars without extensive coronal regions or strong radiative fluxes. Some applications include main sequence stars of spectral types A an later, pre-main sequence stars, and giants and supergiants with effective temperatures in the range $15{,}000 > T_{eff}(K) > 3{,}000$. Hybrid models, in which radiation is considered as a complement to Alvén waves have also been considered. The introduction of Alfvén waves leads to modifications in the momentum equation, which includes a new term corresponding to the Lorentz force, proportional to $\mathbf{J} \times \mathbf{B}$, where $\mathbf{J}$ is the current density an $\mathbf{B}$ is the magnetic field intensity, as we have seen in Chap. 2.

The results of these models show a better fit for main sequence stars, such as the Sun, while for red giants the observed velocities are generally lower than predicted, which makes it necessary to include a relatively uncertain wave damping correction.

10.5   An Estimate of the Mass Loss Rate in Hot Stars

Let us estimate the mass loss rate in a hot star based on a very simple model for the absorption mechanism of stellar radiation by unsaturated spectral lines. Assuming that the gas is accelerated from the photosphere where $r = R_*$ up to a region where $r = R$ and $v = v_f$, we have

$$\dot{M} = 4\pi r^2 \rho(r) v(r) \tag{10.23}$$

at any position r. For $r \gg R_*$, or $r \gg R$, which is equivalent to assuming that $R - R_* \ll r_\infty$, where r_∞ is the total size of the envelope, we have

$$R^2 \rho(R) v_f \simeq r^2 \rho(r) v_f \,,$$

so that

$$R^2 \rho(R) \simeq r^2 \rho(r) \qquad (r \gg R) \,. \tag{10.24}$$

Assuming that the envelope contains only ionized H and He, we have

$$\rho \simeq n_p m_H + 4 n_{He} m_H \simeq n_p \left(1 + 4\frac{n_{He}}{n_p}\right) m_H \simeq n_p (1 + 4y) m_H \tag{10.25}$$

where $y = n_{He}/n_H \simeq n_{He}/n_p$ is the He abundance by number of atoms.

Let k be the line producing element, having an abundance of $a_k \simeq n_k/n_p$, and $Q_{ki} = n_{ki}/n_k$ be the fraction of k atoms in ionization stage i. In these conditions, the mass loss rate (10.23) becomes

$$\dot{M} = 4\pi R^2 \rho(R) v_f \simeq 4\pi R^2 v_f n_p (1 + 4y) m_H \simeq 4\pi R^2 v_f \frac{n_k}{a_k} (1 + 4y) m_H$$

or

$$\dot{M} \simeq 4\,\pi\,R^2\,v_f\,\frac{n_{ki}}{Q_{ki}\,a_k}\,(1+4\,y)\,m_H \ . \tag{10.26}$$

Usually, the ratio $n_{ki}/Q_{ki}\,a_k$ and the proton density depend on position inside the envelope. However, let us assume that these quantities are essentially constant and equal to the their value at $r = R$.

As we have seen, the terminal velocity v_f can be estimated directly from the P Cygni profile. The density of the absorbing particles can be obtained from the line equivalent width W_v, defined by

$$W_v = \int_0^\infty \frac{I_c - I_v}{I_c}\,dv \ , \tag{10.27}$$

where I_v and I_c are the line and continuum intensities, respectively, and the integral covers the whole line. For unsaturated lines, we have

$$I_v \simeq I_c\,e^{-\tau_v} \simeq I_c\,(1 - \tau_v) \ , \tag{10.28}$$

where τ_v is the line optical depth, given by

$$\tau_v \simeq N_{ki}\,\alpha_v \ , \tag{10.29}$$

where N_{ki} is the column density of the absorbing ions and α_v is the absorption cross section. Therefore, the equivalent width W_v can be written as

$$W_v \simeq \int_0^\infty \tau_v\,dv \simeq \int_0^\infty N_{ki}\,\alpha_v\,dv \simeq N_{ki}\int_0^\infty \alpha_v\,dv$$

$$W_v \simeq N_{ki}\,\frac{\pi\,e^2}{m_e\,c}\,f \ , \tag{10.30}$$

where we have introduced the line oscillator force f. In terms of the wavelength λ,

$$W_\lambda \simeq \int_0^\infty N_{ki}\,\alpha_\lambda\,d\lambda = \frac{\lambda^2}{c}\,W_v \ , \tag{10.31}$$

that is,

$$W_\lambda \simeq \frac{\pi\,e^2\,\lambda^2}{m_e\,c^2}\,N_{ki}\,f \ . \tag{10.32}$$

We can estimate the column density N_{ki} integrating $n_{ki}(r)$ in the envelope, approximately from $r = R$ to $r = r_\infty$,

$$N_{ki} \simeq \int_R^{r_\infty} n_{ki}(r)\, dr \simeq \int_R^{r_\infty} n_{ki}(R)\, \frac{R^2}{r^2}\, dr \simeq n_{ki}(R)\, R^2 \int_R^{r_\infty} \frac{dr}{r^2}$$

$$N_{ki} \simeq n_{ki}(R)\, R^2 \left(\frac{1}{R} - \frac{1}{r_\infty} \right), \qquad (10.33)$$

where we have used (10.24). For $r_\infty \gg R$,

$$N_{ki} \simeq n_{ki}(R)\, R . \qquad (10.34)$$

Substituting (10.32) and (10.34) into (10.26), we get

$$\dot{M} \simeq 4\,\pi\, R\, v_f\, \frac{W_\lambda\, m_e\, c^2}{\pi\, e^2\, \lambda^2\, f\, Q_{ki}\, a_k}\, (1 + 4\,y)\, m_H . \qquad (10.35)$$

Naturally, this model is extremely crude, and does not take into account the effect of the gas expansion on the radiation absorption process, or the ionization equilibrium in the envelope. However, using typical values for the absorption lines, we can obtain mass loss rates similar to the observations. For example, let us consider the Si IV $\lambda 1{,}393$ Å line in a O-type star, having the following parameters: $R \simeq R_* \simeq 30\,R_\odot$, $v_f \simeq 1{,}500$ km/s, $W_\lambda \simeq 5$ Å, $\lambda \simeq 1{,}393$ Å, $f \simeq 0.5$, $Q_{ki} \simeq 2 \times 10^{-3}$, $a_k \simeq 3 \times 10^{-5}$ and $y \simeq 0.1$. Using Eq. (10.35) we get $\dot{M} \simeq 1.4 \times 10^{-6}\, M_\odot$/year, which is similar to the observed rates in hot stars.

10.6 An Estimate of the Mass Loss Rate in Cool Stars

Analogously to the previous section, let us estimate the mass loss rate in a cool giant star using a very simple model to account for the action of the stellar radiation pressure on grains. The rate can be written as

$$\dot{M} \simeq 4\,\pi\, r^2\, \rho\, v \simeq 4\,\pi\, r^2\, n\, m\, v , \qquad (10.36)$$

where $n = \rho/m$ is the number density and m the average mass of the gas particles. The force per unit volume on the gas can be written as

$$n\, m\, \frac{dv}{dt} \simeq \frac{L_*}{4\,\pi\, r^2\, c}\, Q\, n_d\, \sigma_d \qquad (10.37)$$

where L_* is the stellar luminosity, Q is the grain efficiency factor, n_d is the grain density and σ_d is the grain absorption cross section. The grain optical depth is

$$d\tau_d \simeq n_d\, \sigma_d\, Q\, dr \qquad (10.38)$$

Using (10.36)–(10.38) the mass loss rate can be written as

$$\dot{M} \frac{dv}{d\tau_d} \simeq \dot{M} \frac{dv}{dt} \frac{dt}{dr} \frac{dr}{d\tau_d} \simeq (4\pi r^2 n m v) \frac{dv}{dt} \frac{1}{v} \left(\frac{1}{n_d \sigma_d Q} \right)$$

$$\dot{M} \frac{dv}{d\tau_d} \simeq \left(n m \frac{dv}{dt} \right) \left(\frac{4\pi r^2}{n_d \sigma_d Q} \right) \simeq \left(\frac{L_*}{4\pi r^2 c} Q n_d \sigma_d \right) \left(\frac{4\pi r^2}{n_d \sigma_d Q} \right) ,$$

that is,

$$\dot{M} \frac{dv}{d\tau_d} \simeq \frac{L_*}{c} . \tag{10.39}$$

Adopting a very rough approximation, $dv/d\tau_d \simeq v_f/\tau_d$, where τ_d is the total optical depth, we have

$$\frac{\dot{M} v_f}{\tau_d} \simeq \frac{L_*}{c} \tag{10.40}$$

or

$$\dot{M} \simeq \frac{L_* \tau_d}{c v_f} . \tag{10.41}$$

Equation (10.41) can be compared to (10.13), which corresponds to $\tau_d \simeq 1$. We can estimate the optical depth τ_d considering that

$$\tau_d \simeq n_d \sigma_d Q R \simeq \frac{n_d}{n} n \pi a^2 Q R , \tag{10.42}$$

where R represents again the base of the envelope, n is the gas density, and $\sigma_d \simeq \pi a^2$ is essentially the geometric cross section of a spherical grain with radius a. We have then

$$\tau_d \simeq \frac{\rho_d}{\rho} \frac{m_H}{m_d} n \pi a^2 Q R \simeq \frac{\rho_d}{\rho} \frac{\pi a^2 m_H n Q R}{\frac{4}{3} \pi a^3 s_d} ,$$

that is

$$\tau_d \simeq \frac{3 m_H}{4} \frac{n Q R}{a s_d} \frac{\rho_d}{\rho} , \tag{10.43}$$

where m_d is the grain mass, s_d is the grain internal density and ρ_d/ρ is the grain-gas ratio, which is exactly the inverse of the gas-dust ratio ρ/ρ_d we have seen in Sect. 2.5.7. Using typical values for a circumstellar envelope around a red giant star, $n \simeq 10^8 \text{ cm}^{-3}$, $Q \simeq 1$, $R \simeq R_* \sim 10^{14} \text{ cm}$, $a \simeq 1{,}000 \text{ Å}$, $s_d \simeq 3 \text{ g/cm}^3$ and

$\rho_d/\rho \simeq 1/200$, we get $\tau_d \simeq 2$, not very different from the value $\tau_d \simeq 1$ mentioned above. In fact, if $\tau_d \ll 1$ the grains are not efficient in the momentum transfer to the gas. If $\tau_d \gg 1$, the radiation is completely absorbed and therefore is not observed, so that $\tau_d \simeq 1$ is an average that satisfies both restrictions. Finally, applying (10.41) to a giant star such as Mira, with $\tau_d \simeq 1$, $L_* \sim 10^3 L_\odot$ and $v_f \simeq 10$ km/s, we get $\dot{M} \simeq 2 \times 10^{-6} M_\odot$/year. Similar values can be obtained using Reimers formula (10.7), taking $R_* \simeq 10^3 R_\odot$, $M_* \simeq 1 M_\odot$ and $\eta \simeq 3$.

10.7   Interaction with the Interstellar Medium

The interaction of the stellar winds with the interstellar medium proceeds as the envelope expands, leading to the heating, ionization and energy and momentum transfer to the neighbouring regions, particularly by shock waves (Chap. 9). Classic examples are the interaction of the solar wind with the local interstellar medium, the HII regions, associated with hot, young stars, and planetary nebulae, associated with intermediate mass stars in their late evolutionary stages. In the last examples, the gas is ionized by the ultraviolet photons from the central star, and simultaneous dynamic effects are produced by the expansion of the envelope, forming "bubbles" or "superbubbles" that are observed in the Milky Way and in other spiral galaxies.

The chemical composition of the interstellar gas is also changed by the contamination with the heavy elements produced by the stellar nucleosynthesis processes – both quiescent and explosive – and also by the expulsion from the stars of solid grains coagulated in the cool atmospheres and envelopes.

The analysis of the propagation of shock waves in the interstellar medium is frequently highly idealized, in agreement with the hypotheses made in Chap. 9, such as the assumption of isothermal or adiabatic shocks. The regions affected by the shock usually have three different phases, according to the propagation stage of the shock. In the first phase the wind expands freely, then follows an adiabatic phase, and then a phase called *snowplow*. The latter is generally more important due to its longer duration and by the observational effects that can be predicted. In this phase the interstellar medium reaches the most external part of the shock, is heated and cools rapidly to a relatively low temperature of the order of 10^4 K. The gas mass swept by the wave is larger than the mass of the formed bubble, and remains in the compressed region. This region, denser and relatively cool, is pushed by the wave like a snowplow, with a duration up to about 10^6 year, depending on the wind intensity.

The formation theory of bubbles and other dense regions caused by the propagation of shock waves in stellar winds can be applied to several important astrophysical situations. Examples are the formation of planetary nebulae, the ring nebulae around Wolf-Rayet stars, and the nebulae around hot stars of O spectral type still embedded in the molecular clouds where they were born. The first of these processes is the subject of the wind interaction theory leading to the formation of a planetary nebula, as mentioned in Chap. 9. However, the detailed formation process is not well known, and the the origin of the observed asymmetries in many nebulae is still controversial.

10.8   Mass Loss and Stellar Evolution

Mass loss may affect significantly the stellar evolution, particularly for massive stars, in which the mass ejection is important already on the main sequence. For example, a $50\,M_\odot$ star with a mass loss rate of $\dot{M} \sim 10^{-5}\,M_\odot$/year loses about $20\,M_\odot$ in just 2×10^6 years, a short time scale in terms of stellar evolution. On the other hand, a solar type star needs about 10^{13} years to lose about 10 % of its original mass at a typical rate of $10^{-14}\,M_\odot$/year, so that the mass loss process in these objects is only meaningful in the later evolutionary stages (AGB stars and planetary nebulae), where rates of about 10^{-6}–$10^{-4}\,M_\odot$/year can be observed.

One of the consequences of mass loss by ejection of the stellar photospheric layers is the unveiling of the internal layers, where frequently elements produced by stellar nucleosynthesis are located, having been dredged up from the lower layers by convection. This occurs for example in carbon stars or in Wolf-Rayet stars. Also the younger planetary nebulae (usually bipolar nebulae, or the so-called "type I" objects) present some He and N enrichment relative to the interstellar composition at the epoch their progenitor stars were formed.

The stellar luminosity, and consequently the evolution timescales, are also affected by mass loss. A massive star with a high mass loss rate during the He-burning phase reaches lower masses as it evolves towards the giant branch, so that it also reaches lower luminosities. Therefore, its evolution time scale is longer than it would be without mass loss, since about half of the original mass may be lost by the stellar winds.

The evolution of *very* massive stars is usually considered on the basis of HR diagram tracks up to the supernova or black hole stage. Mass loss affects these tracks, as mentioned, and also the "onion shell" chemical composition structure. The changes depend on the mass loss rate, which depends on the stellar position on the HR diagram. On the main sequence the mass loss rate is of the order of $10^{-6}\,M_\odot$/year, and may increase by one order of magnitude during the evolution towards the giant or supergiant branch. During the LBV (luminous blue variable) phase, rates up to $10^{-4}\,M_\odot$/year may be reached in short time scales of about 10^4 years. During the Wolf-Rayet phase, rates of the order of $10^{-5}\,M_\odot$/year are observed with time scales of about 10^6 years. Supernova explosions occur in a scale of the order of a few times 10^6 years, when about half of the original mass would have been lost, that is, up to about 50 % of the initial stellar mass is ejected in the final explosive phase, including the mass of the collapsed remnant.

In the case of intermediate mass stars, as we have seen, the evolutionary tracks are affected in a significant way only during the late stages. The more massive of these stars have two or three convective dredge up processes, enriching the gas in He and N, and decreasing the C abundance. The time scales are longer, of the order or higher than 10^8 years. At the AGB stage, the mass loss rates increase by several orders of magnitude, from about 10^{-14} to $10^{-10}\,M_\odot$/year on the main sequence to about 10^{-7} to $10^{-5}\,M_\odot$/year at the final stage. For example, a star with $1\,M_\odot$ on the main sequence reaches the giant branch with practically the same mass. At this point, with a mass loss rate of the order of 10^{-6} $M_\odot$/year, the star loses about 0.1 $M_\odot$ in 10^5 years. At the superwind phase the mass loss rate is higher, of the order of 10^{-5} $M_\odot$/year or higher, and the star loses approximately 0.2 $M_\odot$ in 2×10^4 years, typically the age of a planetary nebula with mass of about 0.2 $M_\odot$. The remnant mass is about $m_r \simeq 1.0 - 0.1 - 0.2 \simeq 0.7$ $M_\odot$, essentially the observed mass of white dwarf stars.

Exercises

10.1. By the application of the radiation driven wind theory to hot stars it is possible to derive a correlation between the modified wind momentum, $\dot{M} \, v_f \, \sqrt{R_*}$, and the stellar luminosity L_* given by

$$ \log(\dot{M} \, v_f \, \sqrt{R_*/R_\odot}) \simeq -1.37 + 2.07 \, \log(L_*/10^6 L_\odot) , $$

where the mass loss rate is in $M_\odot$/year, and the terminal velocity v_f is in km/s. The modified wind momentum depends weakly on the stellar mass, so that this dependency is not included in the relation above. The B0Ia star ϵ Ori has an effective temperature $T_{eff} = 28{,}000$ K and radius $R_* = 33$ $R_\odot$. It presents an intense wind, with terminal velocity of $v_f = 1{,}500$ km/s estimated from P Cygni profiles. (a) Estimate the mass loss rate. (b) Assume that the terminal velocity is reached at $r \simeq 2\,R_*$. What is the circumstellar gas density in this region?

10.2. Show that the expression in the right member of (10.35) has dimensions of mass per unit time. How would this equation be modified by replacing the line equivalent width by W_ν, the corresponding quantity per unit frequency?

10.3. Show that the action of the stellar radiation pressure due to an optically thick line corresponds to a force per unit mass proportional to $v \, (dv/dr)$. Consider the limit of large velocity gradients, where the line width is essentially due to the Doppler shift caused by the wind velocity. Hint: Consider that the radiative momentum absorbed per cm^3 per second is proportional to $F_\nu(r) \, \Delta v/c$, where $F_\nu(r)$ is the monochromatic flux at r and Δv is the line Doppler width corresponding to the velocity gradient at a distance of 1 cm.

10.4. The M1.5I star α Sco has a mass $M_* = 18$ $M_\odot$, radius $R_* = 600$ $R_\odot$ and luminosity $\log(L_*/L_\odot) = 4.6$. The wind terminal velocity is $v_f = 20$ km/s. (a) Determine the mass loss rate using the approximate expression (10.41), in which the mass loss is due to the dust, and also using Reimers formula, Eq. (10.7). (b) Considering that the rate obtained by more accurate methods is $dM/dt = 1.0 \times 10^{-6}$ $M_\odot$/year, what is the total grain optical depth τ_d? What is the value of the efficiency factor η in Reimers formula?

10.5. In a coronal wind, part of the energy can be transferred by electron conduction. The conduction flux ($\mathrm{erg\,cm^{-2}\,s^{-1}}$) is $F_c = -\kappa_c\,(dT/dr)$, where κ_c is the thermal conductivity coefficient, given by $\kappa_c = \kappa_0\,T^{5/2}$, where $\kappa_0 = 1.0 \times 10^{-6}\,\mathrm{erg\,cm^{-1}\,s^{-1}\,K^{-7/2}}$. (a) The conductive luminosity L_c (erg/s) is the total energy transferred by conduction per second through a sphere of radius r. How must the temperature change with position r to keep the conductive luminosity constant? (b) In a model for the solar corona, the temperature decreases with position according to the table below. What is the value of the conductive luminosity in this model? How does the luminosity L_c change with position r? What fraction of the solar luminosity is transferred by conduction?

$r(R_\odot)$	1.0	5.0	10.0	20.0	30.0	40.0	50.0
$T(10^6\,\mathrm{K})$	2.0	1.15	0.90	0.75	0.70	0.65	0.60

Bibliography

Castor, J.I., Abott, D.C., Klein, R.I.: Radiation-driven winds in of stars. Astrophys. J. **195**, 157 (1975) (Fundamental paper on stellar winds, particularly by the treatment of the radiative acceleration in the point source limit)

Clayton, D.D.: Principles of Stellar Evolution and Nucleosynthesis. University of Chicago Press, Chicago (1984) (New printing with a new preface of a classic text on stellar evolution and nucleosynthesis in an advanced level. Originally published in 1968)

Habing, H.J.: Circumstellar envelopes and asymptotic giant branch stars. Astron. Astrophys. Rev. **7**, 97 (1996) (Comprehensive review paper on mass loss in cool stars and wind models. See also Lafon, J.-P., Berruyer, N., Astron. Astrophys. Rev. **2**, 249 (1991))

Habing, H.J., Olofsson, H.: Asymptotic Giant Branch Stars. Springer, New York (2004) (A comprehensive collection of articles by experts on all aspects of the physics of AGB stars)

Holzer, T.E., Axford, W.: The theory of stellar winds and related flows. Ann. Rev. Astron. Astrophys. **8**, 31 (1970) (Excellent review article on the stellar wind theory, containing the equations of isothermal winds and energy transfer)

Iping, R.C., Sonneborn, G.: FUSE observations of mass loss in planetary nebulae. In: IAU Symposium 209, 187, Ann Arbor, (2003) (An account of ultraviolet observations in several galactic planetary nebulae. Figure 10.3 is based on data from this reference. See also Iping, R.C. et al.: IAU Symposium 234, 429 (2006), Sonneborn, G.: IAU Symposium 209, 405 (2003) and Sonneborn, G. et al.: IAU Symposium 234, 513 (2006))

Keller, G.R., Herald, J.E., Bianchi, L., Maciel, W.J., Bohlin, R.C.: A new grid of synthetic spectra for the analysis of [WC]-type central stars of planetary nebulae. Monthly Not. Roy. Astron. Soc. **418**, 705 (2011) (A study of stellar winds in hydrogen-deficient central stars of planetary nebulae, presenting a new numerical grid of detailed stellar atmosphere models with applications to some real stars. Figures 10.4 and 10.5 are from this reference)

Lamers, H.J.G.L.M., Cassinelli, J.P.: Introduction to Stellar Winds. Cambridge University Press, Cambridge (1999) (Referred to in Chapter 1. Includes a general discussion on stellar winds throughout the HR diagram)

Lucy, L.B., Solomon, P.M.: Mass loss by hot stars. Astrophys. J. **159**, 879 (1970) (Pioneer article on the momentum transfer from the radiation field in hot stars to the expanding gas based on line radiation absorption)

Maciel, W.J., Keller, G.R., Costa, R.D.D.: Metallicity effects on the modified wind momentum of CSPN. Rev. Mexicana Astron. Astrof. **44**, 221 (2008) (A study of the modified wind momentum in the central stars of planetary nebulae and the effect of the metallicity. See also Pauldrach, A.W.A. et al., Astron. Astrophys. **419**, 1111 (2004) and Kudritzki, R.P., Puls, J., Ann. Rev. Astron. Astrophys. **38**, 613 (2000))

Mihalas, D.: Stellar Atmospheres. Freeman, San Francisco (1978) (Referred to in Chap. 1. Includes two chapters on line formation in moving media and mass loss through stellar winds in the framework of stellar atmospheres)

Mihalas, D., Weibel-Mihalas, B.: Foundations of Radiation Hydrodynamics. Dover, New York (1999) (Second edition of a text originally published in 1984, with an advanced treatment of radiation hydrodynamics and astrophysical applications)

Owocki, S.: Stellar winds. Planets, stars and stellar systems In: Oswalt, T.D., Barstow, M.A. (eds.) Stellar Structure and Evolution, vol. 4, p. 735. Springer, New York (2013) (Recent comprehensive review paper on stellar winds)

Parker, E.: Dynamics of the interplanetary gas and magnetic fields. Astrophys. J. **128**, 664 (1958) (The original solution of the isothermal wind equations applied to the solar wind is due to Eugene Parker. See also Astrophys. J. **132**, 821 (1960); Astrophys. J. **143**, 32 (1966) and Space Sci. Rev. **4**, 666 (1965))

Sobolev, V. V.: Moving Envelopes of Stars. Harvard University Press, Cambridge (1960) (Classic book on the line formation theory and radiative transfer in expanding atmospheres)

Spitzer, L.: Physical Processes in the Interstellar Medium. Wiley, New York (1978) ([Student edition: 1998] Referred to in Chapter 9. Basic text on the physical processes in the interstellar medium with a good discussion on the interaction of expanding nebulae and the interstellar medium)

Solutions

Problems of Chapter 1

1.1 We have $P \simeq 1$ atm $\simeq 1.0 \times 10^6$ dyn/cm^2 and $T \simeq 300$ K. Since $P \simeq nkT$, $n \simeq 2.4 \times 10^{19}$ cm^{-3}, so that $\lambda \sim n^{-1/3} \sim 3.5 \times 10^{-7}$ cm. The dimensions are about $R \sim 3$ m $\simeq 300$ cm, then $\lambda \ll R$.

1.2 Using the same procedure of Sect. 1.3,

$\frac{\partial}{\partial t} \int_V \mu dV = - \oint \mu \mathbf{v} \cdot \mathbf{n} dS + \int_V Q dV,$

so that $\frac{\partial}{\partial t} \int_V \mu dV + \int_V \mathbf{\nabla} \cdot (\mu \mathbf{v}) dV = \int_V Q dV,$

and $\int_V \left[\frac{\partial \mu}{\partial t} + \mathbf{\nabla} \cdot (\mu \mathbf{v}) \right] dV = \int_V Q dV,$

or $\frac{\partial \mu}{\partial t} + \mathbf{\nabla} \cdot (\mu \, \mathbf{v}) = Q.$

1.3 Since $\frac{dM}{dt}$ (g/s) $= 4\pi r^2$ (cm^2) ρ (g/cm^3) v (cm/s), we have

$\frac{dM}{dt}$ ($M_\odot$/year) $= \frac{dM}{dt}$ (g/s) $\frac{3.16 \times 10^7}{1.99 \times 10^{33}} \frac{\text{s}}{\text{year}} \frac{M_\odot}{\text{g}}$ and

$\frac{dM}{dt}$ ($M_\odot$/year) $\simeq 1.6 \times 10^{-26} \frac{dM}{dt}$ (g/s);

$\frac{dM}{dt}$ ($M_\odot$/year) $\simeq 2.0 \times 10^{-25} \; r^2 \rho \, v$ (cm/s),

$\frac{dM}{dt}$ ($M_\odot$/year) $\simeq 2.0 \times 10^{-20} \; r^2 \rho \, v$ (km/s).

1.4 For the Sun, $\dot{M} \simeq 4\pi R_\odot^2 (\rho v)_{R_\odot} = 3.0 \times 10^{-14}$ $M_\odot$/year, so that

$(\rho v)_{R_\odot} = j_{R_\odot} = \frac{3.0 \times 10^{-14}}{1.6 \times 10^{-26}} \frac{1}{4\pi (6.96 \times 10^{10})^2} = 3.1 \times 10^{-11}$ g cm^{-2} s^{-1}.

At 1 AU this must be multiplied by $(R_\odot / 1 AU)^2$, so that

$j_{AU} = 6.7 \times 10^{-16}$ g cm^{-2} s^{-1} or $\frac{j_{R_\odot}}{j_{AU}} = 4.6 \times 10^4$.

1.5

(a) The mass loss rate is $\dot{M}$ ($M_\odot$/year) $= 2.0 \times 10^{-20} \, r^2$(cm^2) ρ(g/cm^3) v(km/s). The stellar luminosity is $\log(L/L_\odot) = 3.2$ or $L = 1.58 \times 10^3 L_\odot = 6.10 \times 10^{36}$ erg/s. Since $L = 4\pi R^2 \sigma T_{ef}^4$ the stellar radius is $R = 1.3 \times 10^{11}$ cm $= 1.8 R_\odot$ and $r \simeq 1.3 \times 10^{12}$ cm. The gas density is then $\rho \simeq 9.9 \times 10^{-14}$ g/cm^3.

W.J. Maciel, *Hydrodynamics and Stellar Winds: An Introduction*, Undergraduate Lecture Notes in Physics, DOI 10.1007/978-3-319-04328-9, © Springer International Publishing Switzerland 2014

(b) Since $\rho = n\mu m_H$ and

$$\mu m_H = \frac{n_H m_H + n_{He} m_{He}}{n_H + n_{He}} = \frac{1 + 4(n_{He}/n_H)}{1 + (n_{He}/n_H)} m_H$$

we have $\mu = \frac{1 + 4 \times 0.1}{1 + 0.1} = 1.27 \simeq 1.3$ and $n = \frac{9.9 \times 10^{-14}}{1.3(1.67 \times 10^{-24})} \simeq 4.6 \times 10^{10}$ cm^{-3}.

This is valid for neutral winds. For ionized winds, μ decreases and n increases.

Problems of Chapter 2

2.1 Instead of the integral leading to Eq. (2.86) we have

$\int_0^{v(R)} v\,dv = -\int_{r_0}^R g_* dr = -\int_{r_0}^R \frac{GM_*}{r^2} dr$, which can be integrated as

$$\frac{v(R)^2}{2} = -GM_* \int_{r_0}^R r^{-2} dr = GM_* \left(\frac{1}{R} - \frac{1}{r_0} \right)$$

and $v(R)^2 = 2GM_* \left(\frac{1}{R} - \frac{1}{r_0} \right) = 2g_*(R)R\left(1 - \frac{R}{r_0}\right)$.

Taking again $r_0 = 2R$, for g_* constant (2.86) gives $v(R) = 62$ km/s, so that $v(R)^2 = 2g_*(R)R(1 - 1/2) = g_*(R)R = \frac{1}{2}(62)^2$ and $v(R) = \frac{62}{\sqrt{2}} \simeq 44$ km/s.

Assuming $v(r) = \bar{v} = v(r = 1.5R)$, with $r = (3/2)R$, we have

$$\bar{v}^2 = 2GM_* \left(\frac{1}{r} - \frac{1}{r_0} \right) = 2GM_* \left(\frac{2}{3R} - \frac{1}{2R} \right) = \frac{GM_*}{3R}, \text{ then } \bar{v} = 25.2 \text{ km/s}$$

and $t \simeq \frac{R}{\bar{v}} \simeq 32$ days. More correctly, we have $t = 47.4$ days.

2.2 From the hydrostatic equilibrium equation, $\frac{dP}{dr} = -\frac{GM_*\rho}{r^2}$, so that

$$\left.\frac{dP}{dr}\right]_\odot \simeq -\frac{GM_\odot \bar{\rho}_\odot}{R_\odot^2} = -\frac{GM_\odot}{R_\odot^2} \frac{M_\odot}{(4/3)\pi R_\odot^3} \simeq -\frac{3GM_\odot^2}{4\pi R_\odot^5} \simeq -4 \times 10^4 \text{ dyn/cm}^3.$$

Comparing with the model, $\frac{(dP/dr)_{mod}}{(dP/dr)} \simeq 3.3$.

2.3

(a) $T(1\ AU) = 1.5 \times 10^6 \left(\frac{6.96 \times 10^{10}}{1.5 \times 10^{13}} \right)^{2/7} = 3.2 \times 10^5 \text{ K} = 0.22\ T_0.$

(b) Using Example 2.5.2, $\frac{dP}{dr} = -\frac{GM_\odot n m_H}{r^2}$ and $P = 2nkT$, so that

$\frac{dP}{dr} = 2kT\frac{dn}{dr} + 2nk\frac{dT}{dr}$ and

$$\frac{dT}{dr} = T_0 \frac{2}{7} \left(\frac{r_0}{r} \right)^{-5/7} \left(-\frac{r_0}{r^2} \right) = -\frac{2}{7} \frac{T_0 r_0}{r^2} \left(\frac{r}{r_0} \right)^{5/7}.$$

We have then

$2kT\frac{dn}{dr} + 2nk\frac{dT}{dr} = -\frac{GM_\odot n m_H}{r^2}$, or

$$2kT_0 \left(\frac{r_0}{r} \right)^{2/7} \frac{dn}{dr} = (2nk)\frac{2}{7} \frac{T_0 r_0}{r^2} \left(\frac{r}{r_0} \right)^{5/7} - \frac{GM_\odot n m_H}{r^2}.$$

The density is

$$\frac{dn}{n} = \left[\frac{2k}{7r^2} \frac{2}{2k} \frac{T_0}{T_0} \frac{r_0}{r_0} \left(\frac{r}{r_0} \right)^{5/7} \left(\frac{r}{r_0} \right)^{2/7} - \frac{GM_\odot m_H}{r^2 2kT_0} \left(\frac{r}{r_0} \right)^{2/7} \right] dr,$$

which can be written as

$$\frac{dn}{n} = \left[\frac{2}{7r} - \frac{1}{H}\left(\frac{r}{r_0}\right)^{-12/7}\right] dr, \text{ where } \frac{1}{H} = \frac{GM_\odot m_H}{2kT_0 r_0^2}.$$

Integrating from n_0 to n and from r_0 to r, we have

$$\ln n = \ln n_0 + \frac{2}{7}\ln\left(\frac{r}{r_0}\right) - \frac{7r_0}{5H}\left[1 - \left(\frac{r_0}{r}\right)^{5/7}\right], \text{ or}$$

$$n = n_0 \left(\frac{r}{r_0}\right)^{2/7} \exp\left[-7\, r_0[1 - (r_0/r)^{5/7}]/5H\right] \text{ and } n(1 \ AU) = 4.9 \times 10^4 \text{ cm}^{-3}.$$

2.4

(a) $v_0^2 = \frac{kT_{ef}}{\mu m_H} = 5.51 \times 10^{12}$, so that $v_0 = 23.5$ km/s. Using the relation given, we have $\frac{R}{R_*} \simeq 0.997$.

(b) Using (2.92), $v(r) = 23.5 + (2{,}500 - 23.5)\left(1 - \frac{R/R_*}{r/R_*}\right)^{0.8}$, and

$$(0.60)(2{,}500) = 1{,}500 = 23.5 + (2{,}500 - 23.5)\left(1 - \frac{0.997}{r_{60}/R_*}\right)^{0.8},$$

so that $\frac{r_{60}}{R_*} = 2.09$. The plot is shown below.

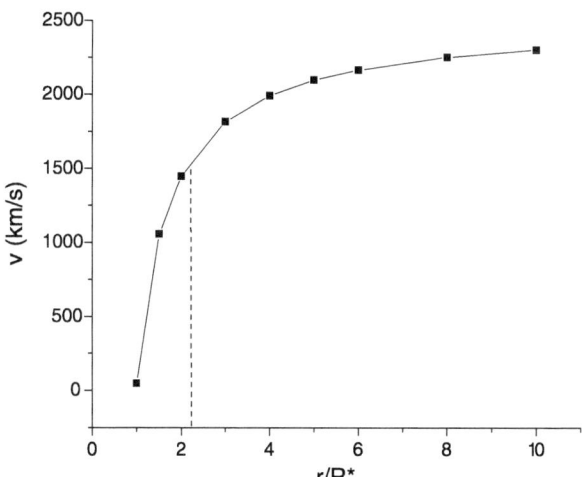

2.5

(a) $\frac{v_f}{v_e} = 2.6$, so that $v_e = \frac{v_f}{2.6} = \frac{2{,}200}{2.6} = 846$ km/s.

The escape velocity is given by $v_e = \sqrt{\frac{2(1-\Gamma_e)GM_*}{R_*}}$ so that $\Gamma_e = 1 - \frac{v_e^2 R_*}{2GM_*}$.

With the stellar radius and mass $\Gamma_e = 0.37$.

(b) Since $\Gamma_e = \frac{\kappa_e L_*}{4\pi c GM_*}$, we have $\kappa_e = \frac{4\pi c GM_* \Gamma_e}{L_*} = 0.36$ cm²/g.

Problems of Chapter 3

3.1 The gas pressure on the container walls can be written as
$P = \frac{1}{3}\int_0^\infty p\,v(p)\,n(p)\,dp,$
where $n(p)dp$ is the density of particles with momentum between p and $p + dp$ and $v(p)$ is the particle velocity. In the case of a non-relativistic perfect gas, we can use the Maxwell-Boltzmann distribution function, so that the momentum distribution can be written as
$n(p)dp = \frac{4\pi np^2 dp}{(2\pi m_e kT)^{3/2}}\exp(-p^2/2m_e kT),$
where n is the total particle density and $m_e = 9.11\times 10^{-28}$ g is the electron mass. Therefore, the equation of state can be written as
$$P = nkT\left[\frac{4}{3\sqrt{\pi}}\int_0^\infty x^{3/2}e^{-x}dx\right],$$
where $x = p^2/(2m_e kT)$. The term within brackets is equal to the unity, so that the equation reduces to (3.8).

3.2

(a) $n_{HeII}/n_{HII} = 0.10$ and $n_e = n_{HII} + n_{HeII}$ so that
$$\mu = \frac{1}{m_H}\frac{n_{HII}m_H + n_e m_e + n_{HeII}m_{He}}{n_{HII} + n_e + n_{HeII}} \simeq \frac{1 + 4(n_{HeII}/n_{HII})}{2(1 + n_{HeII}/n_{HII})} = \frac{1.4}{2.2} = 0.636.$$
(b) $n_{HeIII}/n_{HII} = 0.10$ and $n_e = n_{HII} + 2n_{HeII}$ so that
$$\mu = \frac{1}{m_H}\frac{n_{HII}m_H + n_e m_e + n_{HeIII}m_{He}}{n_{HII} + n_e + n_{HeIII}} \simeq \frac{1+4(n_{HeIII}/n_{HII})}{2+3(n_{HeIII}/n_{HII})} = \frac{1.4}{2.3} = 0.609.$$

3.3

(a) $\rho = n\mu m_H = (10^4)(0.64)(1.67\times 10^{-24}) = 1.07\times 10^{-20}$ g/cm^3. The equation of state gives $P = nkT = (10^4)(1.38\times 10^{-16})(10^4) = 1.38\times 10^{-8}$ dyn/cm^2.
(b) $\rho v = (1.07\times 10^{-20})(20\times 10^5) = 2.14\times 10^{-14}$ g cm^{-2} s^{-1}.
(c) $\alpha = 30" = 1.45\times 10^{-4}$ rad $= \frac{R}{d}$, so that the radius is $R = (1.45\times 10^{-4})(2,000) = 0.29$ pc. The time scale is given approximately by $\tau \sim \frac{R}{v} \sim$ 0.29 pc $/20$ km/s $= 4.49\times 10^{11}$ s $= 1.42\times 10^4$ year.

3.4

(a) From the equation of state $P = \frac{k\rho T}{\mu m_H}$ we get
$\rho = \frac{(1.3)(1.67\times 10^{-24})(10^{3.5})}{(1.38\times 10^{-16})(3,000)} = 1.66\times 10^{-8}$ g/cm^3, and the particle density is
$n = \frac{\rho}{\mu m_H} = 7.65\times 10^{15}$ cm^{-3}.
(b) $g = \frac{GM}{R^2}$ so that
$R = \sqrt{\frac{GM}{g}} = \sqrt{\frac{(6.67\times 10^{-8})(1.99\times 10^{33})}{10}} = 3.64\times 10^{12}$ cm $= 52R_\odot$. The mass loss rate can be written as $\dot{M} = 4\pi r^2\rho v$, so that
$\rho = \frac{\dot{M}}{4\pi r^2 v} = \frac{(10^{-6})(1.99\times 10^{33})}{(3.16\times 10^7)(4\pi)(25)(3.64\times 10^{12})^2(20\times 10^5)} = 7.57\times 10^{-15}$ g/cm^3,
and $n = 3.49\times 10^9$ cm^{-3}. The pressure is then
$P = \frac{k\rho T}{\mu m_H} = \frac{(1.38\times 10^{-16})(7.57\times 10^{-15})(10^3)}{(1.3)(1.67\times 10^{-24})} = 4.81\times 10^{-4}$ dyn/cm^2

Comparing with the pressure at the stellar atmosphere, we have
$\frac{P_*}{P} = \frac{3.16 \times 10^3}{4.81 \times 10^{-4}} = 6.57 \times 10^6$.

3.5

(a) In the region considered, $20,000 = 1.5\,T_0$, so that $T_0 = 13,300$ K.
The pressure is
$$P_0 = \frac{k\rho_0 T_0}{\mu m_H} = \frac{(1.38 \times 10^{-16})(2 \times 10^{-11})(13,300)}{1.67 \times 10^{-24}} = 21.98 \text{ dyn/cm}^2, \text{ and}$$
$\log P_0 = 1.34$.
At the region characterized by T_{eff},
$$P = \frac{(1.38 \times 10^{-16})(9 \times 10^{-10})(20,000)}{1.67 \times 10^{-24}} = 1,487.4 \text{ dyn/cm}^2, \text{ and } \log P = 3.17, \text{ so}$$
that $\Delta P = 1,465$ dyn/cm^2.

(b) $\frac{n_e}{n_{e0}} = \frac{\rho}{\rho_0} = 45$, so that $n_{e0} = \frac{\rho_0}{m_H} = 1.20 \times 10^{13} \text{ cm}^{-3}$ and $\log n_{e0} = 13.08$.
Deeper in the atmosphere we have $n_e = \frac{\rho}{m_H} = 5.39 \times 10^{14} \text{ cm}^{-3}$, and $\log n_e = 14.73$, so that $\Delta n = 5.27 \times 10^{14} \text{ cm}^{-3}$.

Problems of Chapter 4

4.1 We have $dQ = dE + P\,dV = \frac{dE}{dT}dT + P\,dV$,
but $c_V = \frac{1}{\nu}\frac{\partial Q}{\partial T}\Big]_V = \frac{1}{\nu}\frac{dE}{dT}$, and $c_P = \frac{1}{\nu}\frac{\partial Q}{\partial T}\Big]_P$.
Using the equation of state $P = \frac{\nu\mathscr{R}T}{V}$ we have
$dQ = \nu c_V dT + \frac{\nu\mathscr{R}T}{V}dV = 0$, $c_V\frac{dT}{T} = -\mathscr{R}\frac{dV}{V}$ and $\frac{dT}{T} + \frac{\mathscr{R}}{c_V}\frac{dV}{V} = 0$, so that
$\gamma = \frac{c_P}{c_V} = 1 + \frac{\mathscr{R}}{c_V}$ and $\frac{\mathscr{R}}{c_V} = \gamma - 1$, from which we get
$\frac{dT}{T} + (\gamma - 1)\frac{dV}{V} = 0$. Since
$P \propto \frac{T}{V}, \frac{dP}{P} = \frac{dT}{T} - \frac{dV}{V}$ or $\frac{dP}{P} = -\gamma\frac{dV}{V}$. Also
$\rho \propto \frac{1}{V}$ and $\frac{d\rho}{\rho} = -\frac{dV}{V}$, that is
$\frac{dP}{P} = \gamma\frac{d\rho}{\rho}$ and $\frac{d\log P}{d\log\rho} = \frac{d\ln P}{d\ln\rho} = \gamma = 5/3$. We have then
$E = \frac{3}{2}\mathscr{R}T, c_V = \frac{3}{2}\mathscr{R}$, and $c_P = c_V + \mathscr{R} = \frac{5}{2}\mathscr{R}$.

4.2 From the equation of state, $PV = \frac{\nu\mathscr{R}T}{\mu}$. If P is constant, $P\,dV = \frac{\nu\mathscr{R}}{\mu}dT$.
From the first law of Thermodynamics,
$dQ = \nu c_V dT + \frac{\nu\mathscr{R}}{\mu}dT$,
where $c_P = \frac{1}{\nu}\frac{\partial Q}{\partial T}\Big]_P = c_V + \frac{\mathscr{R}}{\mu}$, so that $\gamma = \frac{c_P}{c_V} = \frac{c_V + \mathscr{R}/\mu}{c_V} = 1 + \frac{\mathscr{R}}{\mu c_V}$.

4.3 Differentiating (4.30) relative to r and substituting into (4.29) we have
$\frac{dT}{dr} = \frac{T_0}{n_0^{\gamma-1}} (\gamma - 1) n^{\gamma-2}\frac{dn}{dr}$, that can be written as
$$2kT\frac{dn}{dr} + 2nk\left[\frac{T_0}{n_0^{\gamma-1}} (\gamma - 1)n^{\gamma-2}\frac{dn}{dr}\right] = -\frac{GM_\odot nm_H}{r^2}.$$
Grouping the terms in dn/dr, Eq. (4.31) follows. Using $h = \frac{2kT_0 r_0^2}{GM_\odot m_H}$, we get

$$n^{\gamma-2}\,dn = -\frac{GM_\odot m_H n_0^{\gamma-1}}{2\gamma k T_0}\,r^{-2}\,dr = -\frac{n_0^{\gamma-1}}{\gamma h}\left(\frac{r_0}{r}\right)^2\,dr,\text{ which is (4.32).}$$

Integrating from r_0 to r,

$$\int_{n_0}^n n^{\gamma-2}\,dn = -\frac{n_0^{\gamma-1} r_0^2}{\gamma h}\int_{r_0}^r r^{-2}\,dr,\text{ we get}$$

$$\left[\frac{n^{\gamma-1}}{\gamma-1}\right]_{n_0}^n = -\frac{n_0^{\gamma-1} r_0^2}{\gamma h}\left[\frac{r^{-1}}{-1}\right]_{r_0}^r\quad\text{or}$$

$$\frac{1}{\gamma-1}n^{\gamma-1} = \frac{1}{\gamma-1}n_0^{\gamma-1} - \frac{n_0^{\gamma-1} r_0^2}{\gamma h}\left(\frac{1}{r_0}-\frac{1}{r}\right).$$

This can be reduced to (4.33).

4.4 Using Eq. (4.37) we have $h = e + PV^* = e + \frac{P}{\rho}$. Using (4.74) we get

$$\epsilon = \rho\,\mathbf{v}\left(\frac{v^2}{2} + \phi + h\right) + \mathbf{q}.$$

4.5 From the first Eq. (4.66), using (1.15), we get (4.76). Assuming steady state, we obtain Eq. (4.77). Using $r^2\rho v = $ constant, we get the equation

$$4\pi r^2\rho v\left(\frac{v^2}{2} + e + \frac{P}{\rho}\right) - 4\pi r^2 K\frac{dT}{dr} - 4\pi r^2\rho v\frac{GM_*}{r} = \text{constant.}$$

From this relation we obtain (4.78). From the first Eq. (4.66), in analogy to (4.77), we get

$$\frac{d}{dr}\left[r^2\rho v\left(\frac{v^2}{2} + e + \frac{P}{\rho}\right) - r^2 K\frac{dT}{dr}\right] + r^2\rho v\left(\frac{GM_*}{r^2} - g_r\right) = 0.$$

From this relation we have (4.81).

Problems of Chapter 5

5.1 We can write $\mathbf{v}' = M_1\,\mathbf{v}$ and $\mathbf{v}'' = M_2\,\mathbf{v}'$ so that $\mathbf{v}'' = M_2 M_1\mathbf{v} = M_3\,\mathbf{v}$, and $M_3 = M_2 M_1$. We have

$$M_2 = \begin{pmatrix} \cos\phi & 0 & \sin\phi \\ 0 & 1 & 0 \\ -\sin\phi & 0 & \cos\phi \end{pmatrix}$$

and

$$M_1 = \begin{pmatrix} \cos\theta & \sin\theta & 0 \\ -\sin\theta & \cos\theta & 0 \\ 0 & 0 & 1 \end{pmatrix}$$

so that

$$M_3 = \begin{pmatrix} \cos\theta \, \cos\phi & \sin\theta \, \cos\phi & \sin\phi \\ -\sin\theta & \cos\theta & 0 \\ -\cos\theta \, \sin\phi & -\sin\theta \, \sin\phi & \cos\phi \end{pmatrix}.$$

5.2 The number of components in each case is given in the table below.

n	$r = 4$	$r = 10$
0	$4^0 = 1$	$10^0 = 1$
1	$4^1 = 4$	$10^1 = 10$
2	$4^2 = 16$	$10^2 = 100$
3	$4^3 = 64$	$10^3 = 1{,}000$

5.3 We can write the matrix

$$M = \begin{pmatrix} a_{11} & a_{12} & a_{13} \\ a_{21} & a_{22} & a_{23} \\ a_{31} & a_{32} & a_{33} \end{pmatrix} \begin{pmatrix} b_{11} & b_{12} & b_{13} \\ b_{21} & b_{22} & b_{23} \\ b_{31} & b_{32} & b_{33} \end{pmatrix}$$

$$M = \begin{pmatrix} X_{11} & X_{12} & X_{13} \\ X_{21} & X_{22} & X_{23} \\ X_{31} & X_{32} & X_{33} \end{pmatrix}$$

We have the following equations:
$$X_{11} = a_{11}b_{11} + a_{12}b_{21} + a_{13}b_{31}$$
$$X_{12} = a_{11}b_{12} + a_{12}b_{22} + a_{13}b_{32}$$
$$X_{13} = a_{11}b_{13} + a_{12}b_{23} + a_{13}b_{33}$$
$$X_{21} = a_{21}b_{11} + a_{22}b_{21} + a_{23}b_{31}$$
$$X_{22} = a_{21}b_{12} + a_{22}b_{22} + a_{23}b_{32}$$
$$X_{23} = a_{21}b_{13} + a_{22}b_{23} + a_{23}b_{33}$$
$$X_{31} = a_{31}b_{11} + a_{32}b_{21} + a_{33}b_{31}$$
$$X_{32} = a_{31}b_{12} + a_{32}b_{22} + a_{33}b_{32}$$
$$X_{33} = a_{31}b_{13} + a_{32}b_{23} + a_{33}b_{33}$$

$$\begin{pmatrix} A'_{11} & A'_{12} & A'_{13} \\ A'_{21} & A'_{22} & A'_{23} \\ A'_{31} & A'_{32} & A'_{33} \end{pmatrix} = M \begin{pmatrix} A_{11} & A_{12} & A_{13} \\ A_{21} & A_{22} & A_{23} \\ A_{31} & A_{32} & A_{33} \end{pmatrix}$$

$$A'_{11} = (a_{11}b_{11}+a_{12}b_{21}+a_{13}b_{31})A_{11}+(a_{11}b_{12}+a_{12}b_{22}+a_{13}b_{32})A_{21}+(a_{11}b_{13}+a_{12}b_{23}+a_{13}b_{33})A_{31} \quad \text{etc.}$$

5.4 Nonzero solutions

i	j	k
1	2	3
1	3	2
2	1	3
2	3	1
3	1	2
3	2	1

$(\mathbf{A} \times \mathbf{B})_i = \epsilon_{ijk} A_j B_k$

$= \epsilon_{123} A_j B_k + \epsilon_{132} A_j B_k + \epsilon_{213} A_j B_k + \epsilon_{231} A_j B_k + \epsilon_{312} A_j B_k + \epsilon_{321} A_j B_k$

$= \epsilon_{123} A_2 B_3 + \epsilon_{132} A_3 B_2 + \epsilon_{213} A_1 B_3 + \epsilon_{231} A_3 B_1 + \epsilon_{312} A_1 B_2 + \epsilon_{321} A_2 B_1$

$= A_2 B_3 - A_3 B_2 - A_1 B_3 + A_3 B_1 + A_1 B_2 - A_2 B_1$

$= A_y B_z - A_z B_y - A_x B_z + A_z B_x + A_x B_y - A_y B_x.$

5.5

(a) From (5.21), $(\nabla f)_x = \frac{\partial f}{\partial x} = 2x - 2y = 2(x - y)$; $(\nabla f)_y = \frac{\partial f}{\partial y} = -2x + 2y = -2(x - y)$; $(\nabla f)_z = \frac{\partial f}{\partial z} = 4z.$

(b) From (5.22), $\nabla \cdot \mathbf{v} = \frac{\partial v_x}{\partial x} + \frac{\partial v_y}{\partial y} + \frac{\partial v_z}{\partial z} = 2 - 2 + 2z - 2z = 0.$

Problems of Chapter 6

6.1 From the continuity equation in vector form we can write $\nabla \cdot (\rho \mathbf{v}) = 0$, so that $\rho \nabla \cdot \mathbf{v} + (\mathbf{v} \cdot \nabla)\rho = 0$ or $\rho \nabla \cdot \mathbf{v} = 0$ and $\nabla \cdot \mathbf{v} = 0$, so that we have $\frac{\partial v_k}{\partial x_k} = 0.$

6.2

(a) The rotation matrix is

$$M = \begin{pmatrix} \cos\theta & \sin\theta \\ -\sin\theta & \cos\theta \end{pmatrix}$$

(b) If vector $\mathbf{v}$ follows the vector transformation rules we must have $\mathbf{v}' = M\mathbf{v}$ where M is given above and

$$v' = \begin{pmatrix} \cos\theta & \sin\theta \\ -\sin\theta & \cos\theta \end{pmatrix} \begin{pmatrix} v_x \\ v_y \end{pmatrix} = \begin{pmatrix} v_x \cos\theta + v_y \sin\theta \\ -v_x \sin\theta + v_y \cos\theta \end{pmatrix}$$

$v'_x = v_x \cos\theta + v_y \sin\theta$

$v'_y = -v_x \sin\theta + v_y \cos\theta$ (1)

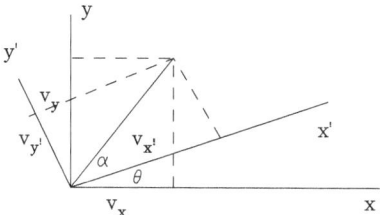

From the figure above we have
$v \cos \alpha = v_x, \; v \sin\alpha = v_y,$
$v \cos(\alpha - \theta) = v'_x, \; v \sin(\alpha - \theta) = v'_y$
$v'_x = v \cos\alpha \cos\theta + v \sin\alpha \sin\theta = v_x \cos\theta + v_y \sin\theta = v'_x$ in (1)
$v'_y = v \sin\alpha \cos\theta - v \cos\alpha \sin\theta = v_y \cos\theta - v_x \sin\theta = v'_y$ in (1).

6.3 If Π_{ik} is a tensor we must have $\Pi_{ik} = P\delta_{ik} + \rho v_i v_k$ and $\Pi'_{ik} = M\Pi_{ik}$ where

$$M = \begin{pmatrix} \cos\theta & \sin\theta \\ -\sin\theta & \cos\theta \end{pmatrix}$$

so that

$$\Pi_{ik} = \begin{pmatrix} P + \rho v_x^2 & \rho v_x v_y \\ \rho v_x v_y & P + \rho v_y^2 \end{pmatrix}$$

$$\Pi'_{ik} = \begin{pmatrix} \cos\theta & \sin\theta \\ -\sin\theta & \cos\theta \end{pmatrix} \begin{pmatrix} P + \rho v_x^2 & \rho v_x v_y \\ \rho v_x v_y & P + \rho v_y^2 \end{pmatrix} = \begin{pmatrix} \Pi'_{11} & \Pi'_{12} \\ \Pi'_{21} & \Pi'_{22} \end{pmatrix}.$$

We have then
$\Pi'_{11} = \cos\theta(P + \rho v_x^2) + \sin\theta\rho v_x v_y$
$\Pi'_{12} = \cos\theta\rho v_x v_y + \sin\theta(P + \rho v_y^2)$
$\Pi'_{21} = -\sin\theta(P + \rho v_x^2) + \cos\theta\rho v_x v_y$
$\Pi'_{22} = -\sin\theta\rho v_x v_y + \cos\theta(P + \rho v_y^2)$
In analogy with Exercise 6.2,
$a'_{ik} = P\delta'_{ik} + \rho v'_i v'_k$
$a'_{11} = P + \rho v'_x v'_y = P + \rho(v_x \cos\theta + v_y \sin\theta)(v_y \cos\theta - v_x \sin\theta)$
or
$a_{11} = P + \rho v_x^2, \; a_{12} = \rho v_x v_y, \; a_{21} = \rho v_x v_y, \; a_{22} = P + \rho v_y^2$
$v'_x = v_x \cos\theta + v_y \sin\theta$
$a'_{11} = a_{11} \cos\theta + a_{21} \sin\theta = (P + \rho v_x^2) \cos\theta + \rho v_x v_y \sin\theta = \Pi'_{11}$
$a'_{12} = a_{12} \cos\theta + a_{22} \sin\theta = \rho v_x v_y + (P + \rho v_y^2) \text{sen}\theta = \Pi'_{12}$
$v'_y = v_y \cos\theta - v_x \sin\theta$
$a'_{21} = a_{21} \cos\theta - a_{11} \sin\theta = \rho v_x v_y \cos\theta - (P + \rho v_x^2) \sin\theta = \Pi'_{21}$
$a'_{22} = a_{22} \cos\theta - a_{12} \sin\theta = (P + \rho v_y^2) \cos\theta - \rho v_x v_y \sin\theta = \Pi'_{22}.$

6.4

(a)

$$F_k = \begin{pmatrix} F_x \\ F_y \\ F_z \end{pmatrix} \text{ and } \delta_{ik} = \begin{pmatrix} 1\ 0\ 0 \\ 0\ 1\ 0 \\ 0\ 0\ 1 \end{pmatrix}$$

$$F_k' = \delta_{ik} F_k$$

$$F_k' = \begin{pmatrix} 1\ 0\ 0 \\ 0\ 1\ 0 \\ 0\ 0\ 1 \end{pmatrix} \begin{pmatrix} F_x \\ F_y \\ F_z \end{pmatrix} = \begin{pmatrix} F_x \\ F_y \\ F_z \end{pmatrix}$$

$$F_k' = F_k = F_i.$$

(b) $\Pi_{i\ell} = \delta_{ik} \Pi_{k\ell}$

$$\Pi_{k\ell}' = \delta_{ik} \Pi_{k\ell}$$

$$\Pi_{k\ell}' = \begin{pmatrix} 1\ 0\ 0 \\ 0\ 1\ 0 \\ 0\ 0\ 1 \end{pmatrix} \begin{pmatrix} \Pi_{11}\ \Pi_{12}\ \Pi_{13} \\ \Pi_{21}\ \Pi_{22}\ \Pi_{23} \\ \Pi_{31}\ \Pi_{32}\ \Pi_{33} \end{pmatrix} = \begin{pmatrix} \Pi_{11}'\ \Pi_{12}'\ \Pi_{13}' \\ \Pi_{21}'\ \Pi_{22}'\ \Pi_{23}' \\ \Pi_{31}'\ \Pi_{32}'\ \Pi_{33}' \end{pmatrix}$$

$$\Pi_{11}' = \Pi_{11}, \Pi_{12}' = \Pi_{12}, \Pi_{13}' = \Pi_{13}$$
$$\Pi_{21}' = \Pi_{21}, \Pi_{22}' = \Pi_{22}, \Pi_{23}' = \Pi_{23}$$
$$\Pi_{31}' = \Pi_{31}, \Pi_{32}' = \Pi_{32}, \Pi_{33}' = \Pi_{33}$$
$$\Pi_{k\ell}' = \Pi_{k\ell} = \Pi_{i\ell}.$$

6.5 The momentum flux tensor for particles of type s is

$\Pi_{ik}^s = (n_s k T)\,\delta_{ik} + m_s n_s v_i v_k = P_s \delta_{ik} + m_s n_s v_i v_k.$

The total momentum flux tensor is

$\Pi_{ik} = \left(\sum_s P_s\right)\delta_{ik} + \left(\sum_s m_s n_s\right) v_i v_k,$ that is,

$\Pi_{ik} = P\delta_{ik} + \rho v_i v_k,$ where P_s is the partial pressure of species s and P is the total gas total pressure.

Problems of Chapter 7

7.1 From (7.9) we have

$\sigma_{ik} = \eta \left(\frac{\partial v_i}{\partial x_k} + \frac{\partial v_k}{\partial x_i} \right) + \lambda \delta_{ik} \frac{\partial v_r}{\partial x_r};$

From (7.24)

$\sigma_{ik} = \eta \left(\frac{\partial v_i}{\partial x_k} + \frac{\partial v_k}{\partial x_i} - \frac{2}{3} \delta_{ik} \frac{\partial v_r}{\partial x_r} \right) = \sigma_{ik} = \eta \left(\frac{\partial v_i}{\partial x_k} + \frac{\partial v_k}{\partial x_i} \right) - \frac{2}{3} \eta \delta_{ik} \frac{\partial v_r}{\partial x_r}.$

Equating these relations, we have $\lambda = -\frac{2}{3}\eta$.

7.2 From the relation given,

$$\sigma_{ik} = \eta\left(\frac{\partial v_i}{\partial x_k} + \frac{\partial v_k}{\partial x_i}\right) - \frac{2}{3}\eta\delta_{ik}\frac{\partial v_r}{\partial x_r} + \xi\delta_{ik}\frac{\partial v_r}{\partial x_r} = \eta\left(\frac{\partial v_i}{\partial x_k} + \frac{\partial v_k}{\partial x_i}\right) - \left(\frac{2}{3}\eta - \xi\right)\delta_{ik}\frac{\partial v_r}{\partial x_r}$$

Comparing with Eq. (7.9),

$$\eta = \eta$$
$$\lambda = -\left(\frac{2}{3}\eta - \xi\right) = \xi - \frac{2}{3}\eta \quad \text{or}$$
$$\xi = \lambda + \frac{2}{3}\eta.$$

7.3 Comparing (7.31) and (7.38) with (7.41),

$$\rho\frac{\partial v_i}{\partial t} = -\frac{\partial \Pi_{ik}}{\partial x_k} + \eta\frac{\partial^2 v_i}{\partial x_k \partial x_k} + F_i, \quad \rho\frac{Dv_i}{Dt} = -\frac{\partial(P\delta_{ik})}{\partial x_k} + \eta\frac{\partial^2 v_i}{\partial x_k \partial x_k} + F_i, \text{ and}$$

$$\rho\frac{Dv_i}{Dt} = -\frac{\partial(\Pi\delta_{ik})}{\partial x_k} + \eta\frac{\partial^2 v_i}{\partial x_k \partial x_k} + \frac{\eta}{3}\frac{\partial^2 v_r}{\partial x_k \partial x_r} + F_i. \text{ We get}$$

$$\rho\frac{\partial v_i}{\partial t} = -\frac{\partial \Pi_{ik}}{\partial x_k} + \eta\frac{\partial^2 v_i}{\partial x_k \partial x_k} + \frac{\eta}{3}\frac{\partial^2 v_r}{\partial x_k \partial x_r} + F_i.$$

7.4 $[v] = L/T \qquad [\rho] = M/L^3 \qquad [\eta] = M/LT.$

v and η depend on T:

$$\frac{v}{\eta} = \frac{LLT}{TM} = \frac{L^2}{M}, \quad \frac{\eta}{v} = \frac{M}{L^2};$$

ρ depends on M:

$$\frac{v\rho}{\eta} = \frac{L}{T}\frac{M}{L^3}\frac{LT}{M} = \frac{M}{L^3}\frac{L^2}{M} = \frac{1}{L}, \quad \frac{\eta}{v\rho} = L;$$

Including L

$$\frac{vL\rho}{\eta} = [non-dimensional] = R$$

$$\frac{\eta}{vL\rho} = [non-dimensional] = \frac{1}{R}.$$

7.5 With the viscosity tensor σ_{ik} (dyn/cm^2), the energy lost by friction per cm^3 per second is $v_i\partial\sigma_{ik}/\partial x_k$. We have

$$\frac{\partial(v_i\sigma_{ik})}{\partial x_k} = v_i\frac{\partial\sigma_{ik}}{\partial x_k} + \sigma_{ik}\frac{\partial v_i}{\partial x_k}.$$

Equation (6.29) becomes

$$\frac{\partial[\rho(v^2/2+e)]}{\partial t} + \frac{\partial\epsilon_k}{x_k} - \left[\frac{\partial(v_i\sigma_{ik})}{\partial x_k} - \sigma_{ik}\frac{\partial v_i}{\partial x_k}\right] = v_i F_i$$

$$\frac{\partial[\rho(v^2/2+e)]}{\partial t} + \frac{\partial\epsilon_k}{x_k} - \frac{\partial(v_i\sigma_{ik})}{\partial x_k} + \sigma_{ik}\frac{\partial v_i}{\partial x_k} = v_i F_i$$

$$\frac{\partial[\rho(v^2/2+e)]}{\partial t} + \frac{\partial(\epsilon_k - v_i\sigma_{ik})}{x_k} + \sigma_{ik}\frac{\partial v_i}{\partial x_k} = v_i F_i.$$

Problems of Chapter 8

8.1 If $\mathbf{v} = \nabla\varphi$, then $\nabla\times\mathbf{v} = \nabla\times(\nabla\varphi) = 0$, that is $\nabla\times\mathbf{v} = 0$ and the flow is non-rotational.

8.2

(a) $v = c_s = \sqrt{\frac{\gamma kT}{\mu m_H}} = \sqrt{\frac{(5/3)(1.38\times10^{-16})(1,000)}{(1.2)(1.67\times10^{-24})}}$ or $v = 3.4$ km/s.

(b) $v_f = 10$ km/s and $f = \frac{3.4}{10} = 34\%$.

8.3 Differentiating (8.35) we get (8.36); Differentiating (8.32) we get (8.37). Substituting (8.37) and (8.36) in (8.33) we get (8.38). Using again (8.16) we have (8.39). Equation (8.38) becomes

$$v\frac{dv}{dr} - \frac{c_s^2}{v}\frac{dv}{dr} - \frac{2c_s^2}{r} + \frac{GM_*}{r^2} = 0$$

$$\frac{dv}{dr}\left(v - \frac{c_s^2}{v}\right) = \frac{2c_s^2}{r} - \frac{GM_*}{r^2}$$

$$\frac{1}{v}\frac{dv}{dr} = \frac{\frac{2c_s^2}{r} - \frac{GM_*}{r^2}}{v^2 - c_s^2} \quad \text{which is (8.40). We have also}$$

$$\frac{r}{v}\frac{dv}{dr} = \frac{d\ln v}{d\ln r} = \frac{\Delta}{v^2 - c_s^2}$$

which is (8.41) and $\Delta = 2c_s^2 - \frac{GM_*}{r}$ which is (8.42).

8.4 For the escape velocity, $E_c = E_p$ and $\frac{1}{2}v^2 = \frac{GM_*}{r}$.

At the critical point we have $v_e(r_c) = \sqrt{\frac{2GM_*}{r_c}}$.

But also $r_c = \frac{GM_*}{2c_s^2}$ and $v(r_c) = c_s = \sqrt{\frac{GM_*}{2r_c}}$

so that $v(r_c) = \frac{1}{2}v_e(r_c)$.

8.5 $\dfrac{1}{v}\dfrac{dv}{dr} = \dfrac{\frac{2c_s^2}{r} - \frac{GM_*}{r^2}}{v^2 - c_s^2}$

$$\lim_{r \to r_c} \frac{1}{v}\frac{dv}{dr} = \frac{1}{c_s}\left(\frac{dv}{dr}\right)_{r_c} = \frac{-\frac{2c_s^2}{r_c^2} + \frac{2GM_*}{r_c^3}}{2c_s\left(\frac{dv}{dr}\right)_{r_c}}$$

At the critical point $r_c = \frac{GM_*}{2c_s^2}$

$$\left(\frac{dv}{dr}\right)_{r_c}^2 = -\frac{c_s^2}{r_c^2} + \frac{GM_*}{r_c^3} = -\frac{c_s^2 4c_s^4}{(GM_*)^2} + \frac{(GM_*)8c_s^6}{(GM_*)^3}$$

$$= \frac{8c_s^6}{(GM_*)^2} - \frac{4c_s^6}{(GM_*)^2} = \frac{4c_s^6}{(GM_*)^2}$$

so that $\left(\frac{dv}{dr}\right)_{r_c} = \frac{2c_s^3}{GM_*}$.

Problems of Chapter 9

9.1 From Eq. (1.6) we have $\frac{\partial \rho}{\partial t} + u\frac{\partial \rho}{\partial x} + \rho\frac{\partial u}{\partial x} = 0$

but $\frac{\partial \rho}{\partial t} = 0$ so that $u\frac{\partial \rho}{\partial x} + \rho\frac{\partial u}{\partial x} = \frac{\partial(\rho u)}{\partial x} = 0$,

that is $\rho u = $ constant.

9.2 From the definition of M_0 we have

$$M_0^2 = \frac{u_0^2}{c_s^2} = \frac{1}{\gamma}\frac{u_0^2 \rho_0}{P_0} \quad (1)$$

From (9.3) we have

$$\frac{P_1}{P_0} = 1 + \frac{\rho_0 u_0^2}{P_0} - \frac{\rho_1 u_1^2}{P_0} \quad (2)$$

Using (1), (2), and (9.1) we get (9.16). Rewriting (9.5)–(9.7)

$$\epsilon_0 = \rho_0 u_0 \left(\tfrac{1}{2} u_0^2 + \tfrac{1}{\gamma-1} \tfrac{P_0}{\rho_0} \right) \quad (3)$$

$$\epsilon_1 = \rho_1 u_1 \left(\tfrac{1}{2} u_1^2 + \tfrac{1}{\gamma-1} \tfrac{P_1}{\rho_1} \right) \quad (4)$$

$$\epsilon_0 + P_0 u_0 = \epsilon_1 + P_1 u_1 \quad (5).$$

From (9.1), (3)–(5) we get

$$\tfrac{1}{2} u_0^2 + \tfrac{\gamma}{\gamma-1} \tfrac{P_0}{\rho_0} = \tfrac{1}{2} u_1^2 + \tfrac{\gamma}{\gamma-1} \tfrac{P_1}{\rho_1} \quad (6) \text{ which is equivalent to (9.8). Using now (9.1),}$$

(1) and (6) we get (9.17). With $\gamma = 5/3$ (9.6) follows.

9.3

(a) Substituting (9.15) in (9.16), we get

$$2 \left\{ \frac{\rho_0}{\rho_1} \left[1 + \gamma M_0^2 \left(1 - \frac{\rho_0}{\rho_1} \right) \right] - 1 \right\} = (\gamma - 1) M_0^2 \left[1 - \left(\frac{\rho_0}{\rho_1} \right)^2 \right].$$

After some algebra, we obtain

$$\frac{\rho_1}{\rho_0} = \frac{(\gamma + 1) M_0^2}{(\gamma - 1) M_0^2 + 2}. \quad (1)$$

Substituting in (9.15), we get

$$\frac{P_1}{P_0} = 1 + \gamma M_0^2 \left[1 - \frac{(\gamma - 1) M_0^2 + 2}{(\gamma + 1) M_0^2} \right].$$

Also after some algebra we obtain

$$\frac{P_1}{P_0} = \frac{2\gamma M_0^2 - (\gamma - 1)}{\gamma + 1}. \quad (2)$$

(b) Dividing the second relation of part (a) by the first we have

$$\frac{P_1}{P_0} \frac{\rho_0}{\rho_1} = \frac{[2\gamma M_0^2 - (\gamma - 1)][(\gamma - 1) M_0^2 + 2]}{(\gamma + 1)(\gamma + 1) M_0^2}.$$

But $P \propto \rho T$ and $T \propto \frac{P}{\rho}$, so that

$$\frac{T_1}{T_0} = \frac{[2\gamma M_0^2 - (\gamma - 1)] [(\gamma - 1) M_0^2 + 2]}{M_0^2 (\gamma + 1)^2}.$$

9.4

(a) We have $\rho u = \phi$ (1), $P + \rho u^2 = \zeta$ (2)

$\tfrac{1}{2} u^2 + \tfrac{5}{2} \tfrac{P}{\rho} = \xi$ (3), $\bar{u} = \tfrac{\zeta}{\phi}$ (4), $\eta = \tfrac{u}{\bar{u}}$ (5).

In the adiabatic case: $c_s^2 = \tfrac{5}{3} \tfrac{P}{\rho}$ (6). Using (1), (2), (4)

$$\bar{u} = \frac{\zeta}{\phi} = \frac{P + \rho u^2}{\rho u} \quad (7), \text{ and } \bar{u} \geq u, \eta \leq 1.$$

(7) leads to $P + \rho u^2 = \rho u \bar{u}$

$\rho u^2 - \rho \bar{u} u + P = 0$

$u^2 - \bar{u} u + \frac{P}{\rho} = 0$

(6) leads to $u^2 - \bar{u} u + \tfrac{3}{5} c_s^2 = 0$ (8)

$c_s^2 = \tfrac{5}{3} (\bar{u} u - u^2)$ (9).

From (3), (6), (9)

$$\xi = \tfrac{1}{2} u^2 + \tfrac{5}{2} \tfrac{P}{\rho} = \tfrac{1}{2} u^2 + \tfrac{5}{2} (\bar{u} u - u^2)$$

$\xi = u(\tfrac{5}{2} \bar{u} - 2u)$ (10).

Using non-dimensional variables, from Eqs. (3) and (10) the total energy per unit mass is

$$e_t = \frac{\xi}{\bar{u}^2} = \frac{u(\frac{5}{2}\bar{u} - 2u)}{\bar{u}^2} = \frac{u}{\bar{u}}(\frac{5}{2} - 2\frac{u}{\bar{u}})$$

$$e_t = \eta(\frac{5}{2} - 2\eta) \quad (11).$$

Internal energy: from (6) and (9)

$$e'_i = \frac{3}{2}\frac{P}{\rho} = \frac{3}{2}u(\bar{u} - u) \quad (12).$$

Using the non-dimensional variable e_i

$$e_i = \frac{e'_i}{\bar{u}^2} = \frac{3}{2}\frac{u}{\bar{u}^2}(\bar{u} - u) = \frac{3}{2}\frac{u}{\bar{u}}(1 - \frac{u}{\bar{u}})$$

$$e_i = \frac{3}{2}\eta(1 - \eta) \quad (13).$$

Kinetic energy,

$$e_k = \frac{1}{2}\frac{u^2}{\bar{u}^2} = \frac{1}{2}\eta^2 \quad (14).$$

Mach number

$$M^2 = \frac{u^2}{c_s^2} = \frac{3u^2}{5u(\bar{u} - u)} \quad (15)$$

$$M^2 = \frac{3}{5}\frac{u}{\bar{u} - u} = \frac{3}{5}\frac{\eta}{1 - \eta} \quad (16).$$

(b) Using (6) and (9)

$$\sigma = \frac{P}{\rho}\frac{\phi^{2/3}}{\rho^{2/3}}\frac{1}{\bar{u}^{8/3}} = u(\bar{u} - u)\frac{\phi^{2/3}}{\rho^{2/3}\bar{u}^{8/3}} \quad (17).$$

Using (1)

$$\sigma = u(\bar{u} - u)\frac{\rho^{2/3}u^{2/3}}{\rho^{2/3}\bar{u}^{8/3}} = \frac{u^{5/3}(\bar{u} - u)}{\bar{u}^{5/3}\bar{u}}$$

$$\sigma = \eta^{5/3}(1 - \eta) \quad (18).$$

Plotting the total energy, the internal energy, the kinetic energy, the Mach number and the entropy as functions of η, it can be seen that, if the total energy is conserved and the entropy increases, the internal energy must increase at the expense of the kinetic energy, and the Mach number decreases.

9.5 From Eqs. (9.2) and (9.4) we get

$$\frac{P_0}{P_1} + \gamma\frac{\rho_1}{\rho_0}\frac{\rho_1 u_1^2}{\gamma P_1} = 1 + \gamma\frac{\rho_1 u_1^2}{\gamma P_1}.$$

But $M_1^2 = \frac{u_1^2 \rho_1}{\gamma P_1}$

$$\frac{P_0}{P_1} + \gamma\frac{\rho_1}{\rho_0}M_1^2 = 1 + \gamma M_1^2$$

and

$$M_1^2 = \frac{1 - P_0/P_1}{\gamma(\rho_1/\rho_0 - 1)}.$$

Substituting in exercise (9.3), $M_1^2 = \frac{A}{B}$, where

$$A = 1 - \frac{\gamma + 1}{2\gamma M_0^2 - (\gamma - 1)}$$

$$B = \gamma\left[\frac{(\gamma + 1)M_0^2}{(\gamma - 1)M_0^2 + 2} - 1\right].$$

This can be simplified to

$$M_1^2 = \frac{M_0^2(\gamma - 1) + 2}{2\gamma M_0^2 - (\gamma - 1)}.$$

If $\rho_1 > \rho_0$ the numerator is < 1;

If $P_1 > P_0$ the denominator is > 1;

We have $\gamma > 1$ and $M_1^2 < 1$.

Problems of Chapter 10

10.1

(a) $L_* = 4\pi R_*^2 \sigma T_{ef}^4$

$L/L_\odot = 6.0 \times 10^5$

$\log(\dot{M} v_f R_*^{1/2}) = -1.37 + 2.07 \log(L_*/10^6) = -1.83$

$\dot{M} \simeq 1.7 \times 10^{-6} M_\odot/\text{year}$

(b) $\dot{M} = 4\pi r^2 \rho v_f$

$$\rho = \frac{\dot{M}}{4\pi r^2 v_f} = \frac{1.7 \times 10^{-6}}{4 \cdot \pi (2 \times 33 \times 6.96 \times 10^{10})^2 (1{,}500 \times 10^5)} \frac{1.9910^{33}}{3.16 \times 10^7}$$

$\rho \simeq 2.7 \times 10^{-15} \text{ g/cm}^3$.

10.2 $[R] = $cm $\quad [v_f] = $cm/s $\quad [W_\lambda] = $cm $\quad [m_e] = $g

$[c^2] = \text{cm}^2/\text{s}^2 \quad [\lambda^2] = 1/\text{cm}^2 \quad [e^2] = \text{s}^2/\text{cm}^3 \text{ g} \quad [m_H] = \text{g}$

$[R v_f W_\lambda m_e c^2 \lambda^2 e^2 m_H] = \text{g/s}$

$W_\lambda = \frac{\lambda^2}{c} W_\nu$

$$\dot{M} = 4\pi R v_f \frac{\lambda^2 W_\nu m_e c^2}{c \pi e^2 \lambda^2 f Q_{ki} a_k} (1 + 4y) m_H$$

$[R] = $cm $\quad [v_f] = $cm/s $\quad [W_\nu] = 1/\text{s} \quad [m_e] = \text{g}$

$[c^2] = \text{cm}^2/\text{s}^2 \quad [e^2] = \text{s}^2/\text{cm}^3 \text{ g} \quad [m_H] = \text{g}$

$[R v_f W_\nu m_e c^2 e^2 m_H] = \text{g/s}$

10.3 The following units apply: [momentum] $= \frac{\text{g cm}}{\text{s}}$

$\left[\frac{\text{momentum}}{\text{cm}^3} \right] = \frac{\text{g}}{\text{cm}^2 \text{ s}}$

$\left[\frac{\text{momentum}}{\text{cm}^3 \text{ s}} \right] = \frac{\text{g}}{\text{cm}^2 \text{s}^2}$

$\frac{\Delta \lambda}{\lambda} = \frac{\Delta v}{v_0} \propto \frac{dv}{dr} \frac{\Delta r}{c} \propto \frac{dv}{dr} \frac{1}{c}$

$\Delta r = 1 \text{cm} \qquad \Delta v \propto \frac{v_0}{c} \frac{dv}{dr} \propto \frac{dv}{dr}$

$\left[\frac{\text{momentum}}{\text{cm}^3 \text{ s}} \right] \propto \left[\frac{\text{force}}{\text{cm}^3} \right] \propto \frac{F_\nu(v) \Delta v}{c}$

Units:

$(\text{erg cm}^{-2} \text{ s}^{-1} \text{ Hz}^{-1}) \quad (\text{Hz cm}^{-1}) \quad (\text{s cm}^{-1})$

$(\text{g cm cm s}^{-2} \text{ cm}^{-2} \text{ s}^{-1} \text{ Hz}^{-1}) \quad (\text{Hz cm}^{-1}) \quad (\text{s cm}^{-1})$

leading to $\left[\frac{\text{g}}{\text{cm}^2 \text{ s}^2} \right]$

$F_\nu(v) \propto r^{-2} \quad \Delta v \propto \frac{dv}{dr}$

$\left[\frac{\text{force}}{\text{volume}} \right] \propto \frac{1}{r^2} \frac{dv}{dr}$

$\left[\frac{\text{force}}{\text{mass}} \right] \propto \frac{1}{\rho} \frac{\text{force}}{\text{volume}} \propto r^{-2} \rho^{-1} \frac{dv}{dr}$

Continuity equation:

$r^2 \rho v = \text{constant}$, so that $v \propto r^{-2} \rho^{-1}$

$\frac{\text{force}}{\text{mass}} \propto v \frac{dv}{dr}$

10.4

(a) $\dot{M} = \frac{L_* \tau_d}{c \, v_f} = \frac{(1.52 \times 10^{38}) \tau_d}{(3 \times 10^{10})(20 \times 10^5)} \frac{1}{6.3 \times 10^{25}}$

$$\dot{M} = 4.03 \times 10^{-5} \, \tau_d \, M_\odot/\text{year}$$
$$\dot{M} = 4 \times 10^{-13} \, \eta \, \frac{(3.98 \times 10^4)(600)}{18} = 5.31 \times 10^{-7} \, \eta \, M_\odot/\text{year}$$

(b) $\dot{M} = 1.0 \times 10^{-6} \, M_\odot/\text{year} \quad \tau_d = 0.025, \eta = 1.9.$

10.5

(a) $L_c = 4\pi r^2 F_c = -4\pi r^2 \kappa_c \dfrac{dT}{dr} = -4\pi r^2 \kappa_0 T^{5/2} \dfrac{dT}{dr}$

L_c constant, $\quad T^{5/2} \dfrac{dT}{dr} \propto r^{-2}$

$T^{5/2} dT \propto \dfrac{dr}{r^2}$ so that $T^{7/2} \propto r^{-1}$ and $T \propto r^{-2/7}$.

If T decreases faster than $r^{-2/7}$, L_c decreases

If T decreases slower than $r^{-2/7}$, L_c increases

(b) Approximate results are shown in the table below.

The fraction of the conductive luminosity is $\dfrac{L_c}{L_\odot} \simeq \dfrac{3 \times 10^{27}}{3.8 \times 10^{33}} \simeq 2 \times 10^{-6}$.

$r(R_\odot)$	$\log r$	$T(10^6 \text{ K})$	$\log T$	dT/dr (K/cm)	L_c (erg/s)
1.0	0.000	2.00	6.301		
3.0	0.477	1.58	6.199	-3.05×10^{-6}	5.24×10^{27}
5.0	0.699	1.15	6.067		
7.5	0.875	1.03	6.013	-7.18×10^{-7}	2.65×10^{27}
10.0	1.000	0.90	5.954		
15.0	1.176	0.83	5.919	-2.16×10^{-7}	1.86×10^{27}
20.0	1.301	0.75	5.875		
25.0	1.398	0.73	5.863	-7.18×10^{-8}	1.24×10^{27}
30.0	1.477	0.70	5.845		
35.0	1.544	0.68	5.833	-7.18×10^{-8}	2.04×10^{27}
40.0	1.602	0.65	5.813		
45.0	1.653	0.63	5.799	-7.18×10^{-8}	2.79×10^{27}
50.0	1.699	0.60	5.778		

Index